D0001877

URBAN MASS TRANSIT

THE LIFE STORY OF A TECHNOLOGY

Robert C. Post

GREENWOOD TECHNOGRAPHIES

GREENWOOD PRESS
Westport, Connecticut • London

Library of Congress Cataloging-in-Publication Data

Post, Robert C.
    Urban mass transit : the life story of a technology / Robert C. Post.
      p. cm. – (Greenwood Technographies, ISSN 1549-7321)
    Includes bibliographical references and index.
    ISBN 0-313-33916-3 (alk. paper)
    1. Urban transportation.   I. Title.
  TA1205.P67   2007
  388.4–dc22   2006028660

British Library Cataloguing in Publication Data is available.

Library of Congress Catalog Card Number: 2006028660
ISBN: 0–313–33916–3
ISSN: 1549–7321

First published in 2007

Greenwood Press, 88 Post Road West, Westport, CT 06881
An imprint of Greenwood Publishing Group, Inc.
www.greenwood.com

Printed in the United States of America

The paper used in this book complies with the
Permanent Paper Standard issued by the National
Information Standards Organization (Z39.48–1984).

10 9 8 7 6 5 4 3 2 1

Every reasonable effort has been made to trace the owners of copyright materials in this
book, but in some instances this has proven impossible. The author and publisher will be
glad to receive information leading to more complete acknowledgments in subsequent
printings of the book and in the meantime extend their apologies for any omissions.

Contents

Series Foreword

In today's world, technology plays an integral role in the daily life of people of all ages. It affects where we live, how we work, how we interact with each other, what we aspire to accomplish. To help students and the general public better understand how technology and society interact, Greenwood has developed *Greenwood Technographies*, a series of short, accessible books that trace the histories of these technologies while documenting *how* these technologies have become so vital to our lives.

Each volume of the *Greenwood Technographies* series tells the biography or "life story" of a particularly important technology. Each "life story" traces the technology, from its "ancestors" (or antecedent technologies) through its early years (either its invention or development) and its rise to prominence to its final decline, obsolescence, or ubiquity. Just as a good biography combines an analysis of an individual's personal life with a description of the subject's impact on the broader world, each volume in the *Greenwood Technographies* series combines a discussion of technical developments with a description of the technology's effect on the broader fabric of society and culture—and vice versa. The technologies covered in the series run the gamut from those that have been around for centuries—firearms and the printed book, for example—to recent inventions that have rapidly taken over the modern world, such as electronics and the computer.

While the emphasis is on a factual discussion of the development of the technology, these books are also fun to read. The history of technology is full of fascinating tales that both entertain and illuminate. The authors—all experts in their fields—make the life story of technology come alive, while also providing readers with a profound understanding of the relationship of science, technology, and society.

Acknowledgments

I am indebted to Kevin Downing at Greenwood Press and Himanshu Abrol at TechBooks for their editorial expertise, to George W. Hilton, John H. White, Jr., and the late George Krambles for inspiration, to Rudi Volti for his critique of my manuscript, and to Dian Post for help and encouragement at every step of the way.

Chronology

1888	Henry Whitney begins electrifying West End Railway, Boston
1890	General Electric begins manufacturing K-type controller
1890	1,262 miles of trolley line in United States
1890	Streetcar ridership in United States, 2 billion
1890	Electric railway line, "the tube," under the Thames, London
1893	Electric streetcars in Chicago
1893	Metropolitan Street [cable] Railway, Broadway, Manhattan
1896	First Budapest underground, from Gisella Place to the Stadtwaldchen Park
1897	First trolley subway, Tremont Street, Boston
1897	Multiple Unit (MU) control system developed by Frank Sprague
1898	Construction begins on first suspended monorail, Wuppertal, Germany
1898	Stephenson factory moves from Manhattan to Elizabeth, New Jersey
1900	First Paris Underground, from Porte Maillot to Porte Vincennes
1900	Brooklyn elevated lines electrified
1902	22,000 miles of trolley lines in United States
1902	Streetcar ridership in United States, 5 billion
1902	Elektrische Hoch und Untergrunde Bahnen opens in Berlin
1903	Interborough Rapid Transit opens in New York
1903	Manhattan elevated lines electrified by August Belmont
1903	J. G. Brill merged with four other streetcar manufacturers
1905	Gasoline-electric buses debut on Fifth Avenue
1907	Street railway companies own 467 "parks and pleasure resorts"
1907	Model T Ford introduced
1908	Storage battery streetcars put into operation on Manhattan
1910	ASRA changes its name to American Electric Railway Association
1910	First regular trackless trolley service, Laurel Canyon, Hollywood, California
1911	First electric starters on automobiles

1912	Cie. J. G. Brill founded in Paris
1912	70,000 streetcars in the United States
1915	62,000 jitneys nationwide
1916	First Birney Safety Car in service, Seattle
1917	Pacific Electric inaugurates bus service, San Bernardino and Redlands
1917	Last horsecars on Manhattan
1918	Trolley lines represent an investment of $4 billion
1919	General Motors Acceptance Corporation (GMAC) formed as GM's credit arm
1920	8.1 million automobiles registered in United States
1921	"Boss" Kettering predicts that "buses are bound to win out in the long run"
1923	Trackless trolleys on Oregon Avenue, Philadelphia
1923	Streetcar ridership in United States, 14 billion
1924	Fageol Safety Coach introduced
1926	J. G. Brill merged with American Car & Foundry as ACF–Brill
1926	Fageol brothers found Twin Coach Company
1929	Electric Railway Presidents' Conference Committee (PCC) formed
1930	Supreme Court rules that streetcar companies are entitled to 7.5 percent return
1930	23.1 million automobiles registered in United States
1931	60 percent of people entering downtown Los Angeles traveling in autos
1932	AERA becomes American Transit Association
1932	San Antonio, Texas, becomes largest United States city to abandon streetcars entirely
1933–1935	New York Railways abandons Manhattan streetcars
1935	First Moscow underground, from Sokolniki Park to Red Square and Krimskaya
1935	Public Utilities Holding Company Act

1936	National City Lines formed as subsidiary of GM, Firestone, and Standard Oil
1936	First PCC cars delivered to Brooklyn & Queens Transit
1939	Yellow Coach diesel-hydraulic buses
1940	Sixth Avenue line, first Manhattan elevated, abandoned
1940	J. G. Brill makes its last trolley cars
1940	2,800 trackless trolleys in operation in North America
1944	National City Lines acquires Los Angeles Railway
1944–1945	Peak years for transit ridership in United States
1946	Metropolitana opens in Rome
1947	Shreveport, Louisiana, transit nearly 100 percent trackless trolley
1949	First PCCs put on second-hand market, by San Diego Electric Railway
1949	Toronto Transit Commission begins work on Yonge Street Subway
1952	4,900 PCCs in operation in North America
1952	6,500 trackless trolleys in operation in North America
1952	Last PCC cars delivered to San Francisco Municipal Railway
1955	Third Avenue Line, last Manhattan elevated, is demolished
1956	President Eisenhower signs Interstate Highway Act
1957	Last streetcars in New York City
1957	Bay Area Rapid Transit District (BART) created
1956–1966	Streetcars abandoned in Dallas, Detroit, Kansas City, Baltimore, Los Angeles, Washington, D.C., and St. Louis
1958	Electric railway removed from San Francisco-Oakland Bay Bridge
1960	Tatra Smichov builds first trams based on PCC designs and patents
1961	First federal loans for urban mass transit
1963	Skokie Swift begins operation
1964	Urban Mass Transportation Administration established
1964	BART construction begins
1965	General Motors settles federal antitrust suit

1968	Washington Metro construction begins
1970	U.S. oil production peaks
1972	BART opens
1973	Streetcars remain in only nine North American cities
1973	U.S. Congress authorizes diverting some Interstate Highway Funds to mass transit
1973	Los Angeles–El Monte busway
1973	Last trackless trolleys in Chicago
1979	Oil consumption worldwide reaches 65 million barrels per day
1981	San Diego Trolley (LRT) begins operation
1981	President Reagan calls the private auto "the last great freedom"
1986	Construction begins on Los Angeles Metro subway
1987	Construction begins on first light rail line in Los Angeles
1988	Judge Doom introduced in *Who Framed Roger Rabbit*
2001	Washington Metro completed
2006	Orange Line busway, San Fernando Valley

Introduction

The automobile, the computer, the railway, the radio—these all bring a distinct image to mind. But the phrase "urban mass transit" imparts a variety of images and the word *transit* has become shorthand for many sorts of conveyances operated by metropolitan or regional "transit authorities." Most commonly they operate buses, usually on public streets but occasionally on reserved busways. In the biggest cities they operate rapid transit or "heavy rail" systems, where electric trains have their own space and can pick up power from a third rail next to the tracks without endangering people to electrocution. Rapid transit systems have various names—the subway, the metro, the underground—and indeed lines are mainly underground, though sometimes they are elevated above city streets or tucked into the median of a superhighway. "Light rail" shares some of the characteristics of rapid transit, but tracks are on the surface and quite often on public streets, and because of this light rail vehicles (LRVs) cannot pick up power from a ground-level third rail in the manner of rapid transit.

At the other end of the spectrum from heavy and light rail is "paratransit," with no fixed guideways and small, maneuverable motor vehicles—reminiscent of what were often called jitneys in the past—that ply routes where the traffic is light. Then, there are more unusual forms of transit such as automated "people movers" and monorails. Monorails have retained a futuristic image ever since the *schwebebahn* (suspended railway) opened in

Wuppertal in the Ruhr district of Germany more than a century ago. As a symbol of modern times, monorails are often seen at theme parks, and theme parks are where people can sometimes ride into the past aboard the earliest kinds of mass transit conveyances, drawn by horsepower. Theme parks are also among the few places where transit is part of a for-profit business. Transit authorities are public agencies that depend on subsidies—that is, direct financial aid from the government. But it is important to bear in mind that urban public transportation began as a private enterprise and people expected it to be profitable, and it was—often extravagantly so. Money changed hands in payment for transit more often than for any other service.

Something else is even more important. Today, the phrase mass transit may bring to mind a variety of images, from subway trains to minibuses. But most important historically—and the main subject of this book—is a type of conveyance that dominated the urban panorama for most of the century between the 1850s and the 1950s and was known by various names but most often as the *streetcar*. Streetcars were especially important in the growth and development of North American cities, but were scarcely less important in other parts of the world (where they were called tramcars or trams): the British Isles, throughout Europe, in Japan, in Australia. In small towns, there might be just one line, a shuttle between the railroad station and the town square. In big cities, streetcar traffic could be overwhelming. Chicago had 2,800 streetcars, London had almost as many (there called trams). From a hotel or office window overlooking Chicago's State Street, one could often see twenty or thirty streetcars coming and going at one time. The same was true of Pennsylvania Avenue in Washington, D.C., Woodward Avenue in Detroit, Market Street in San Francisco, and Canal Street in New Orleans. To handle all the traffic on Market and Canal, there were four tracks. To get out of the way of congestion in Boston and Philadelphia, the tracks dove underground in streetcar subways. At the main intersection in Los Angeles, four different lines running east and west on Seventh Street crossed five north–south lines on Broadway. More than seventy different streetcar lines fanned out from downtown Pittsburgh.

Streetcars were essential to the process of suburbanization, or residential development beyond what was once called "the walking city." Initially they were powered by horses. The first of these horsecars appeared in New York in the 1830s and within 30 years there were horsecars in every American city, and they were beginning to catch on overseas. Then, another sort of streetcar was developed, whose source of propulsion was not readily apparent. The power came from steam engines that might be many blocks or even miles away from the cars out on the line, pulling long cables in slots between the tracks. The first of these cable cars appeared in San Francisco in 1873. While they were never as numerous as horsecars, eventually there were cable cars in cities and towns from coast to coast, and in a few other countries around the world. Their heyday was brief, however, and by the turn of the century they were fast disappearing, as were horsecars. Today, cable cars remain only in the city where they appeared first, San Francisco, and horsecars at Disneyland. Cable cars and horsecars were both superseded by a third kind of streetcar, whose source of propulsion was likewise not apparent, nor was it transmitted by mechanical means. Streetcars with electric motors were put into operation in several cities and towns during the mid-1880s, but it was difficult to master the technical challenges and not until 1888 was there a system that was reasonably trouble-free, in Richmond, Virginia. Electrification caught on most quickly in the United States, where 90 percent of all streetcars had electric motors by the turn of the century.

Power was transmitted to electric streetcars from stationary dynamos at 600 volts or more, which is potentially lethal. Thus, streetcars had to be designed to draw their current from wires that were out of reach. In some of the great cities of Europe, as well as in New York and Washington, D.C., these were located in conduits between the rails, as with cable cars, and contacted through slots in the pavement. But everywhere else they were overhead. The device that picked up current as a streetcar moved along reminded people of a fisherman trolling and was first called a "troller." Soon it became "trolley," and that word came to define the entire conveyance, at least in the United States, and was used interchangeably with the word streetcar. In the movie musical *Meet Me in St. Louis*, Judy Garland sang "The Trolley Song," and Tennessee Williams wrote a play whose backdrop was one of the streets with streetcars in New Orleans: "They told me to take a streetcar named Desire, and then transfer to one called Cemeteries and ride six blocks and get off at Elysian Fields," says Blanche Dubois.

Because this book addresses one particular family of conveyances in detail, it might have been titled more like other technographies as "the trolley" or "the streetcar." But those terms have lost the resonance they once had, streetcar in particular. Rather than being impelled by something external such as horses or cables, an electric streetcar gained traction through its own wheels, and indeed the whole system was often called *traction*. But the same thing was true of an automobile—providing its own traction—and when automobiles started becoming part of everyday life, in the 1920s, people began to attach a different meaning to the word "car." Always before that, a streetcar had been regarded as a public conveyance that ran on a "street railway." Usually there were two sets of tracks in the center of the roadway, to accommodate lines running in each direction, and usually the surface was flush with the pavement so that other traffic could use that part of the road too. People spoke of the "car line," they paid "carfare." While waiting in the street at the "car stop" to catch a ride, they stood in a "streetcar safety zone." Where streetcar tracks turned left or right, there were warning signs: "Watch Out for the Cars." Where streetcars were stored overnight was called the "carbarn." Where they reversed direction was called "the end of the line."

In many countries, but especially in the United States, streetcars were the complex technological device that people came to know best, and they retained that status until automobiles became even more familiar. Streetcars enabled people to commute to work over distances too far to walk, or take a ride out into the countryside, or just go downtown to shop or catch a movie. But familiarity bred contempt. When monetary inflation increased the cost of operating streetcars after World War I, and extravagant profits melted away, people had no sympathy for the men who controlled streetcar companies—"traction magnates" they had been dubbed, or "trolley barons." Although some street railways were planned simply to help with the sale of outlying residential tracts, others were built on the assumption that they would be profitable in and of themselves. People assumed that cities would continue to grow, and that streetcar revenue would always increase because the auto would remain what it was at first, just a "rich man's toy." Beginning in the 1920s, however, more and more Americans could afford to buy autos and treat them as everyday transportation, and streetcar patronage began to decline. Then came the Great Depression of the 1930s and it declined faster. Streetcar lines in small towns would disappear, sometimes with no replacement at all, sometimes by transforming themselves into bus lines, which could operate on public streets without the expense of tracks and wires and powerplants. After a while, the idea of regaining profits by substituting buses for streetcars took hold among the owners of transit systems in bigger towns and then small cities, and then almost every city, no matter how extensive and heavily traveled the system may have been.

This idea also acquired momentum quickly in Great Britain, but nowhere was the change to rubber tires as sweeping as in the United States. A 1912 census counted more than 70,000 streetcars nationwide. Sixty years later, there were less than 1,200 streetcars but mass transit companies were operating 50,000 buses, the great majority with diesel engines. In 1912 there had been streetcars in more than 370 U.S. cities and towns; by the 1970s they remained in operation in only seven: Boston, Philadelphia, Pittsburgh, San Francisco, Cleveland, Newark, and New Orleans. Each of the latter three cities had only one line, none of the others had more than a handful. The largest streetcar system in the Western Hemisphere was in

Canada, in Toronto, and a look at the current edition of the basic reference book, *Jane's Urban Transport Systems*, tells us that Toronto's is still the largest. But it tells much more than that. Even though they are rare or nonexistent in large parts of the world, streetcar (or tramcar) systems remain integral to the urban panorama in several hundred cities. They are about as scarce in France and England as in the United States, and there is only one system in Mexico, one in India, and just a handful in China. But there are nineteen in Japan, twenty-five in Ukraine, nearly sixty in Germany, and more than seventy in Russia, the largest of all in St. Petersburg.

As for the United States, *Jane's* lists the seven cities noted above—nearly all of the lines that survived into the 1970s still survive today—plus more than a dozen in each of two other categories, light rail and "heritage," all of them new since the 1970s. Light rail may be considered as the offspring of a rapid transit father and a streetcar mother. Light rail lines and streetcar lines both use the same kind of vehicles (LRVs), and because their routes almost always entail some street running in "mixed" traffic, they pick up power from wires that are out of harm's way overhead. Light rail is sometimes conceived as a less capital-intensive variety of rapid transit that avoids the enormous expense of digging for subways. For most of their length, light rail lines typically occupy the right-of-way of one-time freight railroads. If there are high-level platforms in stations, the main thing that distinguishes light rail from rapid transit is the overhead wire (Figure I.1). In contrast to LRV lines, heritage lines—those characterized by *Jane's* as being "operated primarily for tourist purposes"—usually have new trackage entirely in city streets, for the whole idea is to recapture the flavor of times past. Although the cars are often technologically modern, they are intended to *look* old (Figure I.2).

Intended to look old: This needs emphasizing. In the United States, transit riders were often appreciative when new buses were substituted for antiquated streetcars in the 1930s, 1940s, and 1950s. But streetcars have been absent from most cities and towns for so long that they are now imbued with an air of nostalgia, and this nostalgia played a significant role in fostering interest in "heritage" lines and even in gaining political support for light rail lines in the 1980s and 1990s. Partly the enthusiasm for light rail is owing to perfectly rational considerations—a sense of electric power being environmentally friendly, a concern about the depletion of finite petroleum reserves. And partly it is because of deep-seated *sentimental* attachments to the traditional trolleys they resemble because they do run in "mixed" traffic and they do pick up power from overhead wires. With light rail, it might be said that one looks to the future by returning to the past. We could call this "a desire named streetcar." The first light rail line to begin operation in the

Figure 1.1: The light rail line connecting Los Angeles and the San Gabriel Valley follows a former Santa Fe Railway right-of-way for almost its entire length; here, in Pasadena, it runs in the median of Interstate 210. There are trains rather than single cars, and passengers step directly aboard from high-level platforms. This is like rapid transit and unlike a typical streetcar line. The major similarity is the power pickup overhead (by means of a double-hinged device called a "pantograph," a modernized version of a trolley pole) whereas rapid transit utilizes an electrified third rail next to the tracks. Such a third rail would not be feasible with this light rail line, or any other, because in places tracks are embedded in public streets and would put people in harm's way. (Lee Furon photo, author's collection)

United States, in 1981, was called the "San Diego Trolley," although San Diego's LRVs are not much like the trolley cars last seen there in 1949. Yet, even urban shuttles and tourmobiles with rubber tires and diesel engines are often *disguised* as old-time streetcars. They are buses, actually, but people call them trolleys because they have been made to resemble old images of a romanticized past.

One other type of bus is also called a trolley, and truly *is* a trolley, though it does not run on rails. This is now called the electric trolley bus (ETB), but was known for many years as a "trackless trolley" and in genealogical terms is a half-brother of the trolley car. ETBs get their current from overhead wires, but have rubber tires and can be steered around obstructions or even driven "off wire" for short distances by means of auxiliary power units (APUs). At one time, there were trackless trolley lines all over Great Britain and much of Europe (where they were known as trolley buses) and in as many

Figure I.2: As part of Lowell National Historical Park, the Park Service operates a trolley line using cars with up-to-date electro-mechanical equipment but mimicking the appearance of streetcars at the turn of the twentieth century. There are similar "heritage" operations in San Jose, Tucson, Dallas, Memphis, Tampa, and elsewhere, all of them intended to appeal to tourists who are attracted to an "old town" ambience. (Author's photo)

as seventy North American cities. But their numbers have now diminished as drastically as have streetcar lines: they are gone from the British Isles, only a handful in France and Germany, two in Mexico, two in Canada, and four in the United States, in Seattle, Boston, Dayton, Ohio—and San Francisco, whose ETB system marks it as a maverick even more so than its cable cars because it is so extensive. San Francisco's 350 ETBs transport 74 million passengers annually on seventeen lines, which, for comparison, exceeds the yearly count for all 156 diesel bus lines operated by the transit authority in Atlanta.

And for perspective on mass transit in the era of globalization, the mechanical components of most of San Francisco's ETBs were manufactured in the Czech Republic. San Francisco's newest LRVs were made by Breda in Bologna, Italy, Philadelphia's by Kawasaki, Boston's by Kinki Sharyo. In the United States, streetcars are an import. It was once quite the opposite, when the John Stephenson Company in New York shipped thousands of streetcars around the world, and, later on, so did the Philadelphia firm of J. G. Brill & Sons, which even opened a branch plant in France. But just as LRV manufacturing has shifted overseas, so, too, the largest remaining street railway networks are overseas; there are tram (and ETB) systems throughout eastern Europe that are as large as the largest American systems were in their heyday. Even so, this book argues that street railways were more significant to the transformation of urban life in the United States than anywhere else, and also that the United States was more often at the forefront of innovation in urban-transit technology.

True, the *motor vehicle* with an internal-combustion gasoline engine was invented in Europe and the first bus line in America, on Fifth Avenue in New York, used double-deckers built by Daimler in England and De Dion-Bouton in France. Furthermore, the first electrified subway line bored under the Thames in London and was called "the tube." But these are exceptions that prove the rule, for many other varieties of mass transit conveyance appeared earlier in the United States than elsewhere, and then they diffused more rapidly. One distinguished urban historian takes this as "an interesting illustration of America's capacity to absorb innovation" (Hall 1998, 757). The railway and the horse-drawn omnibus both had origins overseas, yet it was New Yorkers who got the idea of putting an omnibus on rails and calling it a horsecar. By the time horsecars showed up on the Compagnie de Tramways in Paris in the 1850s, they were part of the fabric of so many American cities that they had became known as *chemin de fer americain*.

As noted, the first trolley-car system that was unarguably successful was in Richmond, Virginia. But this was preceded by many efforts in other American towns and cities that *almost* succeeded, notably in Baltimore, Maryland. The cable car had a tentative tryout in New York before its success in San Francisco. Cable railways eventually operated in twenty-eight U.S. cities, and they diffused overseas as well, to six cities in the British Isles, two in Australia (Melbourne's becoming the largest system in the world), and one each in Lisbon, Paris, and Dunedin, New Zealand. Other American transit innovations diffused more widely. One example is the technology of multiple-unit control (MU), developed on Chicago's West Side Elevated in the late 1890s. MU enables the operator of a rapid-transit train to apply traction through motors geared to the wheels of every car, and it is now used with heavy rail and light rail everywhere. Another example is the standardized high-tech streetcar called the PCC that was developed in the 1930s with funding from the American Transit Association. These were made for only 15 years in the United States and did not really "save the industry" as hoped, and yet U.S. patents and designs were licensed in several countries overseas and one Czech firm eventually manufactured 20,000 tramcars based on the engineering of PCCs, four times as many as were made in America.

In this book, the references will be mostly to *trolleys* rather than *tramcars*, U.S. terminology and U.S. history. Even the author of a book about the electrification of urban transit in Europe at the turn of the twentieth century feels compelled to address the American scene at some length because, as he writes, the trolley "was an American innovation like the horsecar" (McKay 1976, 40). But a focus exclusively on the United States is impossible, as is suggested by the contextual references in this introduction, and one cannot begin *any* narrative of urban history without looking outward. Athens in the fifth century BCE, Florence during the Renaissance, London in the time of Shakespeare—cities of course flourished long before the first settlers arrived in North America from across the Atlantic in the seventeenth century. By then, horse-drawn public conveyances were part of the urban panorama in many parts of the old world.

Yet it happens that the phenomenon of rapid urbanization—the transformation of farmers and peasants into factory workers and city dwellers— has largely taken place in the last two centuries and therefore has been largely coincident with the history of the United States. Many of the first cities to have systematized mass transit were American cities—"systematized" having several essential meanings: fixed routes and scheduled times of operation, designated stops, set fares. This definition would exclude hackney coaches, the forerunners of taxis (hence the expression "hack"), which had become commonplace in European cities by the turn of the nineteenth century. But hackney coaches were "for hire" by the well-to-do, and the fares were expensive. People of modest means had always walked, and the physical extent of cities was limited by the length of time it took to walk across town, typically between where one resided and where one worked in a different neighborhood. Even as industrialization began to alter every aspect of traditional social and economic relationships, and to engage millions of workers in new skilled or unskilled occupations, cities remained "walking cities" and at first only became more crowded.

The advent of steam-powered railways in England and the United States in the 1820s would greatly facilitate interurban travel—that is, travel *between* cities and towns and still other cities. In the expression of the day, steam railways "annihilated space and time" (Schivelbusch 1979, 41). The development of the steam locomotive was what enabled the extension of railroads across settled countryside in the British Isles and Europe, and in the United States fostered new settlement along lines that initially went "from somewhere to nowhere." Yet steam power was not the essence of railroad technology, for there had been railroads, or tram roads or tramways, long before there were steam engines, especially for hauling ore out of mines by means of human muscle power. The essence was the combination of *flanged wheel on iron rail* (the flange guiding the wheels even when the rails curved), a combination that minimized rolling resistance (friction) and also provided an even surface over irregular terrain. This same combination would prevail in urban settings as well, but with steam power rare except where the lines were elevated on long viaducts. On the elevated railways, everything had to be done so much more substantially than just spiking tracks to wooden ties surrounded by cobblestones that the system picked up the descriptive term "heavy rail."

The conveyances that people rode on downtown streets (and above and below the streets as well) would ultimately have their own motors and be propelled by the magic of electricity. The first urban mass transit, however, did not even involve rails. Rather, it was an adaptation to an urban environment of a type of conveyance that steam railroads were putting out of business in the countryside, the overland stagecoach.

As we turn to the biographies of different sorts of mass transit conveyances, first comes the horse-drawn omnibus, then the horsecar on rails, then the cable car, then trolleys of various types. A chapter on the trolley's heyday is followed by a chapter on the challenge posed by the automobile and another chapter on the industry's effort to regain favor by introducing attractive, well-engineered trolleys and—when that failed to work—to cut its losses by switching to rubber-tired vehicles that did not require a separate infrastructure. Finally, there is a brief look at the heavy-rail rapid transit systems which began operating in a sort of parallel universe in a few of the largest cities around the turn of the twentieth century, and at the new wave of mass transit innovation that includes light rail and busways, the funding borne on the wings of government subsidy. It concludes with an afterword on popular perceptions of mass transit's fortunes during the epoch of the trolley, and a notion that gained widespread acceptance after most streetcar systems were gone—that trolleys were actually superior to what replaced them, and that they were the victim of a conspiratorial ploy rather than intrinsic economic weakness or shifting public preferences.

A word about this technography as a whole. In *Notes on the Underground*, the historian Rosalind Williams writes that her purpose was "not to cover the topic, but to uncover it" (Williams 1990, xi). With urban mass transit, the topic does not really need uncovering, as there is a long shelf of fine books already in print, and some that are true classics such as Sam B. Warner's *Streetcar Suburbs*, Charles Cheape's *Moving the Masses*, John McKay's *Tramways and Trolleys*, and George W. Hilton's *The Cable Car in America*. A look at the bibliography will confirm their quantity, and I can attest to quality. My aim in this narrative is to impart an understanding—as it says in the series foreword—"of how technology and society interact." Part of this interaction involves what we have come to know as the "tradeoff." As Hilton puts it, "All forms of economic activity yield external benefits and involve social costs" (Hilton 1982, 15).

Another historian expresses this same idea in different words: "We can never select the one result we want to the exclusion of all others" (Pye 1978, 18). One example will suffice for now. The elevated railway—the first form of mass transit that ran free of interference from regular traffic—debuted above Ninth Avenue in Manhattan in 1870 and within 20 years there were three elevated lines running the length of the island, and a fourth running from South Ferry to Central Park. Guidebooks extolled trains "gliding far overhead in the air ... the daring of science in overcoming difficulties" (quoted in Reed 1978, 71). No doubt, transit was much more rapid on an elevated railway train than on a trolley car that was crowded by all sorts of other traffic. But enthusiasm faded as the "unwanted results" became more irksome: the elevated structure itself blocked out the sun, the trains disturbed sleep in walk-up apartments, and, in short, the social and environmental costs came to be perceived as outweighing the economic benefits.

What follows is built largely around this cost/benefit theme. It is anything but comprehensive or exhaustive—the narrative thread begins in the United States and usually stays there, or in North America anyway. Rather than "covering the topic," it aims only to describe significant interactions of technology and society in the history of urban mass transit. Because most readers will have some familiarity with the subject already, they will bring to the book experience and knowledge of their own. This is as it should be, and I can only repeat another remark in Prof. Williams's *Notes on the Underground*. Each time someone is reminded of some device or some event that "should have been mentioned," I will take this to be measure of success, not failure.

1

Before Electrification

At the turn of the nineteenth century, the population of the United States was 5.3 million, including 850,000 who were held in bondage. This population was almost entirely rural. There were more slaves in the state of Virginia alone, 347,000, than the total number of people residing in what the 1800 census designated "urban territory," 322,000. And, while there was only a slow change in the proportion of citizens who lived in urban territory between 1800 and 1820, after that the proportion increased rapidly as immigrants arrived in evergrowing numbers—in 1820, 8,385 immigrants debarked in Atlantic coast ports; there were 23,322 in 1830; 84,066 in 1840; 369,980 in 1850; and nearly 2.5 million during the decade preceding the Civil War—and evergrowing numbers found employment in cities.

At first, the new city dwellers crowded into the old neighborhoods, living in what became known as tenements. Boston, Philadelphia, and New York actually had greater population densities than London, Paris, and Amsterdam. Many European cities were surrounded by walls dating back hundreds of years, and their residents had usually lived within these walls long after they served any purpose as protective fortifications. American cities were primarily commercial, places where commodities were gathered and distributed, exported and imported, and they were located at the most advantageous sites for ports, often where rivers widened into bays. This meant that deep water interfered with urban expansion as much as walls did in Europe. But eventually some people began regarding congestion as an opportunity to make money by providing transportation beyond the central city. Ferries were established that carried residents across rivers and harbors—from Manhattan to Brooklyn, from Philadelphia to Camden— and real estate developers began to stretch the limits of urban territory.

In the four decades before the Civil War, the total population in urban territory multiplied 900 percent, from 693,000 in 1820 to 6,217,000 in 1860. Boston grew fourfold, New Orleans sixfold, New York eightfold. New York had 123,700 residents in 1820; by 1860 it had 1,174,779. The population of Manhattan, along with Brooklyn, the Bronx, Queens, and Richmond (Staten Island)—all of which would later become part of what we now know as New York City—had once been about equal to that of Philadelphia and its suburbs, but was twice as large by 1860. By then, more people lived in Brooklyn, the Bronx, Queens, and Richmond than in Boston proper and all its suburbs combined, Charlestown, Chelsea, Cambridge, Brighton, Brookline, Roxbury, and Dorchester.

Boston, Philadelphia, and New Orleans were all growing rapidly, as were Baltimore, Charleston, and Savannah, but none of them as rapidly as New York. So, it is not surprising that it was in New York that the omnibus made its debut, as did the rail-borne conveyance that eventually became known as the streetcar because it operated in the street but was initially called the horsecar because of its motive power. Although horsecars were greatly outnumbered by omnibuses until the 1850s, after that they became such a fixture in U.S. cities and towns of all sizes that in Europe, horsecar lines became known as "American railways."

THE OMNIBUS

If the hackney coach was transit-on-demand for the well-to-do, the omnibus was transit on schedule and, literally, "for the many." (The entire word "omnibus" survives today as an adjective meaning *inclusive*, but the last part as a noun; people wait at a "bus stop" for the most ordinary of all mass transit vehicles.) An omnibus had spoked wheels banded with iron tires, and the driver sat ahead of and above the passenger compartment on an open bench, as with a stagecoach. But it differed from a stagecoach in typically having the door at the rear and benches for seating along both sides. The first such conveyance to have the name omnibus began operating in the French city of Nantes in 1826, followed within a year or so by Paris and then London. Shortly after the omnibus first appeared in Nantes, one Abram Brower commissioned a conveyance of the same sort from the Manhattan coach-making firm of Andrew Wade and put it in service on Broadway with the name *Accommodation* painted on the side.

Brower went into business a few months later than the French entrepreneur in Nantes, whose name was Baudry, and it again bears emphasis that the story of urban mass transit is not exclusively an American story, nothing of the sort. But in many important ways the United States was at the forefront of technological change. The author of another technography, *Cars and Culture*, notes that his narrative "centers on developments that have taken place on American soil" even though "the automobile was not an American invention" (Volti 2004, xii–xiii). Likewise, this history of urban mass transit will center on American developments, even though the omnibus was not an American invention. The immediate descendant of the omnibus, the horse-drawn streetcar, *was* an American invention and it was adopted in great numbers more quickly in the United States; so was the electric trolley, and the story of the cable car is very largely an American story. Later, in the 1930s, it was a team of U.S. engineers who developed the first thoroughly modern streetcar built on assembly-line principles, although trams using its patented components were later adopted more enthusiastically in Europe. This was in contrast to the horse-drawn omnibus a century before: Following its introduction in France and England, Europeans reacted tentatively, while it was embraced enthusiastically in America.

The story begins with John Stephenson (1809–1893), who was born in County Armagh in the north of Ireland but came to America with his father and mother when he was an infant. Stephenson was apprenticed to the same craftsman who made the *Accommodation*, Andrew Wade; he opened his own shop at age twenty in 1829; and he sold his first omnibus in 1831 to Wade's erstwhile customer Abram Brower. Omnibuses soon became a familiar sight on the streets of Manhattan. During the 1830s, they also debuted in Philadelphia and Boston, as well as in Dresden, Germany, and during the 1840s in both Baltimore and another German city, Berlin. But they remained most numerous in London, Paris, and New York. As would happen again and again with new modes of urban transit, many people thought that life was better because of the omnibus, and yet nobody was really satisfied. A lot depended on whether the roads they traveled were paved with cobblestones or crushed stone (a process called macadamizing), or perhaps planked, so passage was fairly certain no matter the season. In many cities and almost all small towns, street paving seemed too expensive to be undertaken extensively, and it was not always clear who was responsible even for keeping streets cleaned up. But as an alternative to paving there was already the example of the steam railway—actually, before that, the wheeled trams that had been used in collieries—that is, in the coal-mining industry.

Figure 1.1: Staged in front of the Stephenson factory at Fourth Avenue and Twenty-seventh Street on Manhattan, this omnibus was slated for shipment to Cape Town, South Africa. The dazzling ornamentation was intended to make a statement that this was not, in reality, a conveyance "for all." (From the Smithsonian Institution)

Laying down a wooden substructure of crossties and spiking down wooden stringers capped with strips of iron enabled horses to pull almost twice the load they could handle even when the streets were in good shape.

JOHN STEPHENSON AND THE HORSECAR

For many years, Stephenson continued to build omnibuses for transit operators in the United States and around the world (Figure 1.1). But he is better remembered for having built the first rail-borne horsecar—and rightfully so, for, here, he was not anticipated anywhere else, as he was with his first omnibus for Brower. Stephenson's *John Mason* was ordered in 1832 by (and named after) a bank president who also headed a new railroad company that planned a line from New York—whose northern city limits were then around Thirtieth Street—to the suburb of Harlem, and eventually to the capital at Albany. (It later became part of the New York Central, one of the great "trunk lines" connecting the nation's two largest cities, New York and Chicago.) When the New York & Harlem first opened for business, however, there was only a mile of track running on Fourth Avenue between the Bowery and its terminal at Fourteenth Street, and not until 1836 would it acquire its first steam locomotive.

Although distinct from an omnibus, with doors on the sides and much smaller wheels, the *John Mason* was not a great deal different from cars on other pioneer American railroads that used horse power at the outset, such and the Ponchartrain Railroad between New Orleans and Carrollton and the Baltimore & Ohio's line to Ellicott City, Maryland. One historian calls it "the first streetcar—by accident" (Cudahy 1995, 11). Still, its "first" is deserved, because the New York & Harlem served a purely local function in Manhattan by having designated stops at street corners along the way, as became the practice with all street railways. For the same fare, one could ride two blocks or ten. (Even a short ride might be worth the money if the streets were muddy.) And, after Stephenson delivered three more cars of the same sort to the New York & Harlem, anyone who wanted to catch a ride anytime during the day or evening could be pretty sure one would be coming along. In addition, the company agreed to pave and maintain a roadway between the tracks and for a certain distance on either side, as well as clearing snow. In getting authorization from municipal officials in the form of franchises—that is, a public privilege conferred by a governmental body—most street railway entrepreneurs would be obligated in a similar way. Long afterwards, this would become a significant financial burden, and an increasingly attractive inducement to switch from streetcars to conveyances that did not entail an obligation to take care of the street—the rubber-tired motorbus that became practical for mass transit about a century later.

Before the turn of the twentieth century, the firm founded by Stephenson would build more than 25,000 streetcars for American cities and for export. Until shortly before the Civil War, however, Stephenson specialized in omnibuses and there were hundreds of them on the streets of New York alone. Some of these streets were paved with cobblestones and an omnibus could roll right along, albeit far from quietly. On Fifth Avenue, omnibuses remained a fixture until after the turn of the century, and then the switch was directly to buses with internal combustion engines; there never were overhead trolley wires on Manhattan, and on "Millionaire's Row" there were never tracks either. But Fifth Avenue was probably the best-kept street in the country; the problem with streets nearly everywhere else was that they were barely surfaced at all. This was one of the main reasons why horsecars were adopted more rapidly in America than in England or on the European continent. There, city centers almost always had paving of some sort, but the streets were narrow and twisting. In the United States, streets in newer towns in the Mississippi Valley and around the Great Lakes, and in the newer parts of old seaboard cities, were often comparatively wide and unrestricted, but they often remained "unimproved" even as population doubled and tripled. In Chicago there were planks on State, Madison, and Clark, but during the spring thaw "it could take crews of men working with ropes and planks two days to rescue an omnibus that had slipped . . . into the quagmire" (Young 1998, 36). The engineering of a horsecar line could be pretty rudimentary, but the presence of tracks would almost certainly require that a street be graded and have gutters for drainage.

FLUSH RAILS AND LIGHTWEIGHT CARS

The New York & Harlem's rails along Fourth Avenue had interfered with regular traffic, and coachmen and wagoners regarded them as an annoyance, to say the least. But in the early 1850s a French engineer named Alphonse Loubat began laying out another New York street railway, this one on Sixth Avenue, and his idea was to have the ties buried deeply enough so that the top of the rails were flush with the pavement, which was grooved for flanges. Soon there were horsecar lines on Second, Third, and Eighth avenues with tracks of this same sort that posed minimal interference to other traffic. But the horsecars themselves did not move as easily as they might. Even though any conveyance on rails could be rolled along with the less exertion than an omnibus, horsecars of the *John Mason*-type were still patterned on the design of railroad cars that were pulled by the power of steam and they weighed as much as 4 tons. For animate power, 4 tons was too heavy. When these cars got overloaded, as happened regularly, they were known as "regular horse killers."

The urge to ease their burden was not owing to sensitivities about abuse (the American Society for the Prevention of Cruelty to Animals was not founded until 1866), but rather to diminish the strain on the pocketbooks of the men who owned streetcars: A horse was a valuable commodity. Although Stephenson had concentrated on making omnibuses after his work in the 1830s for the New York & Harlem, he could readily envision a distinct new horsecar architecture, light and less burdensome to horses, and more nimble on crowded city streets. By 1855 he was turning out cars weighing less than 2 tons for lines in New York as well as in Jersey City, Brooklyn, and Boston, and soon he would be exporting these worldwide (Figure 1.2).

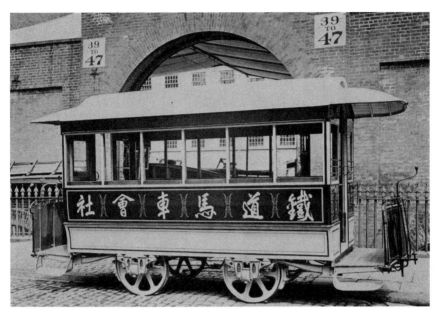

Figure 1.2: A typical Stephenson horsecar of the 1870s or 1880s. This was one of thousands, but unusual in its destination; it was bound for Shanghai and lettered "Company of the Iron Road Horse Car." Stephenson also exported cars to Berlin, Brussels, Birmingham, Copenhagen, Dublin, Lisbon, Paris, St. Petersburg, Tokyo, and Yokohama. (From the Smithsonian Institution)

In the late 1850s, horsecar lines were built in several more eastern cities, notably Baltimore and Philadelphia. For a time, Philadelphia's 148 miles of street railway were more than New York's. There also were lines beyond the Alleghenies, in Cincinnati, Pittsburgh, and St. Louis. Chicago, which had scarcely existed in 1830, had a population of 109,260 in 1860 and it would have more than a million by 1890. (To "grow like Chicago" became an expression heard 'round the world.) Franklin Parmelee (1816–1904) inaugurated omnibus service in Chicago in 1853, and was soon running horsecars on State Street under the auspices of the Chicago City Railway, which would eventually operate one of the most heavily traveled systems in the world. In the far west, there were horsecars in San Francisco before the Civil War and in Los Angeles and San Diego soon after. All told, by 1880 more than 3,000 miles of street railway had been constructed in cities and towns from coast to coast, with nearly 20,000 horsecars in daily operation.

In France, Loubat had introduced "American railways," and an American, George Francis Train, introduced them to England, but not with much success at first because he used the step rail that protruded above the street and interfered with regular traffic. (Pedestrians could "step" over it, but for wheeled vehicles it posed a serious impediment.) Finally, in the 1870s, tramlines began to appear in England with the surface of the rail flush with the pavement, as in Loubat's system. By 1880 trams were also getting well established in German-speaking countries.

BOBTAILS AND DOUBLE-ENDERS

Most smaller cities and towns had "bobtails" with only a step up to the door at the rear. Motive power was provided by a single horse (hence the expression "one-horse town"). Passengers were instructed to put their fare in a box located beside the driver, or into an inclined channel that sent it up to the box by gravity. This device for collecting fares was a patented invention of John B. Slawson, who later joined Stevenson's firm as treasurer. Bobtails were often called "farebox cars" and Booth Tarkington evoked an image of a one-horse streetcar (actually, in Tarkington's Midland it was a mule) in *The Magnificent Ambersons*:

> At the rear door of the car there was no platform, but a step where passengers clung in wet clumps when the weather was bad and the car crowded. The patrons—if not too absentminded—put their fares into a slot; and no conductor paced the heaving floor, but the driver would rap remindingly with his elbow upon the glass of the door to his little open platform if the nickels and the passengers did not appear to coincide in number.

In towns bigger than Midland, or wherever there was considerable traffic, two-horse teams were the norm, as was a two-man crew. The driver was stationed in front, and behind him a conductor actually did "pace the heaving floor," collecting fares and keeping order, just as on a railroad train. He would signal the driver to stop and go by pulling a rope that rang a small bell, while the driver would use a foot pedal to ring a gong as a warning in traffic. A bobtail was "single ended" and had to be turned around at the end of the line: sometimes there was a turntable or the track made a loop; sometimes the body of the horsecar swiveled on its 4-wheel "truck," a design patented by Stephenson. But the larger cars of the sort pictured in Figure 1.2 were "double ended"—where the line terminated, the team was unhitched and led to the other end of the car, and the driver and conductor switched stations. (This would also become the procedure with trolley cars, with the conductor swinging the trolley pole around and the driver—now called the *motorman*—taking his brass handle for the electrical "controller," the device that regulated speed, to the other end.)

Whether bobtails or double enders, horsecars were crucial to the growth of distinct residential neighborhoods. In the United States, a mecca for immigrants, at first these neighborhoods were often populated by the working class and considered less desirable—think about the word *sub*urban along with a word like *substandard*. In later years, however, suburban developments further out toward the periphery attracted office workers and businessmen who wanted to escape noise and congestion, and perhaps put distance between their homes and older neighborhoods that were becoming overcrowded and were often on their way to being termed "slums." As the distances grew longer, however, it was increasingly clear that horsecars left much to be desired—both for the people who had to ride them daily and for those who had invested in them, however profitable they might be.

THE LIMITS OF HORSEPOWER

In discussing any technology it is important to differentiate the self-interest of those who controlled its instruments (owners or managers) from that of those they employed to make them work (labor) and from that of customers who might have their own ideas about how they *should* work. A "consumer" of urban transit wanted to get to his or her destination as quickly as possible—ideally, in comfort, though that remained more an ideal than a reality. Owners of street railways had more money invested in their horses than anything else, as much as $200 each. (Mules cost less, ate less, and worked harder, and they were not uncommon in urban transit, especially with small-scale operations like Booth Tarkington's Midland. But a mule depreciated quickly and could rarely be resold for less arduous work, whereas a secondhand horse could be "farmed out" for half its original value, or even more.) Drivers and conductors represented virtually no investment at all (a day or two of training) and were readily replaceable. They had to expect low wages and extended working hours and, usually, an adversarial relationship with customers. The only shared concern, and for quite different reasons, was with the horses, or, more precisely, with the limitations of horses.

Even with lightweight cars, scenes involving overburdened streetcar horses were a pathetic aspect of urban life. Horses labored on the slightest grade, and even on level ground they strained desperately to get a loaded car rolling from a dead stop. No sooner would it be up to speed than it would be time to slow down for another stop. When their horses were tired, drivers sometimes tried to avoid full stops, and customers had to get on and off moving cars as best they could—one of numerous reasons why they tended to regard drivers as adversaries. Even under ideal conditions, a horsecar could make only 4 or 5 miles per hour, barely faster than a person could walk, and the only advantage of riding was to get out of the weather or avoid teeming sidewalks and filthy gutters. Horses posed a hazard to public health because they dumped large quantities of manure and drenched the pavement with urine. Tetanus was a constant risk. Crowded together in carbarns, horses were susceptible to epidemics, and in 1872, a respiratory and lymphatic disease known as epizootic apthae swept through stables in eastern cities, killing thousands of animals. In New York, so many horses were killed or disabled that oxen were pressed into streetcar service, and the papers ran stories about streetcars being pulled by "gangs of the unemployed."   Smooth, evenly surfaced pavement was desirable for other forms of traffic, for carriages and freight wagons whose drivers wanted to keep moving if at all possible, but not where there were streetcars because horses had trouble gaining traction after a stop unless the street happened to be perfectly dry—which was not often the case, considering the situation with urine. For horsecars, rough cobblestones were better, and that was typically what the owners of streetcar lines used for paving. Streetcar horses could usually be worked for only a few hours each day, and they had a total working life of only a few years, often just 3 or 4 years if they were employed in heavy-duty service in a big city. If a street railway had even moderate hills, the operating company might need as many as ten horses for every horsecar it owned because teams would have to be changed so often, and "helper" teams would be required where hills were steepest. The barns where horses were stabled and cars were stored when not in service might take up acres of valuable real estate. For all sorts of reasons, nobody thought of horse power as a long-term "solution" to the problem of mass transit.

STEAM POWER

It was understood that one potential alternative to horse power was loco-motive (meaning self-propelled) steam power, well proven with steamboats that plied America's rivers and coastal waterways and on the railroads that now stretched to the Pacific. Steam locomotives were usually considered to be an unwelcome intrusion in close quarters on crowded city streets: dirty, noisy, and dangerous because of their tendency to make horses bolt and run (not to mention the possibility of exploding, as a New York & Harlem locomotive did in 1839 at Union Square, killing the crew and injuring several others). Nevertheless, there were many efforts to downscale steam technology so it would be suitable for streetcars. For example, the Baldwin Locomotive Works in Philadelphia turned out small "steam dummies" for service in several cities. Dummies would have their exhaust muffled and their boiler and firebox shrouded in wooden bodies in a seeming attempt to disguise their true nature.

Another way of harnessing steam locomotives in urban transit was to leave them undisguised but get them off the street, up in the air on extended iron viaducts. On such elevated railways, they carried fuel, usually coke, which made less smoke, in a bunker behind the cab and could operate equally well in either direction. Elevateds were erected in several American cities, first on Manhattan and in Brooklyn, and then in Chicago, Boston, and Philadelphia. But the development of urban transit was left largely in the hands of private enterprise—what today we would call the free market—and in most places construction was too capital-intensive to interest investors. Where it was most crowded, however, elevated railways would attract sufficient numbers of paying customers because they made much better time than horsecars: in New York, a trip from South Ferry to Central Park on a Sixth Avenue horsecar took at least an hour; a trip on the "El" was scheduled for 26 minutes. Nevertheless, choices about technology always entails tradeoffs, and tradeoffs are evident throughout the history of urban mass transit—in the case of elevated railways a blight on the cityscape even after electric power began to supersede steam. By the 1890s it began to look like rapid transit could be better provided by tunneling underground, as expensive and disruptive as this might be, and as unsettling as the idea of going underground was to people who did not feel comfortable there.

ANDREW HALLIDIE AND THE CABLE RAILWAY

Rapid transit, important as it was to New York, Boston, Philadelphia, and Chicago—and to a handful of other North American cities when public financing became available in the second half of the twentieth century—is essentially a different branch of the urban-transit family tree, and will be addressed later. There were many technological links between rapid transit and transit on "the surface," however, a notable example being a system of propulsion first tried on an elevated railway with only marginal success that later provided the basis for the first mechanized street railways. As new alternatives to horsepower were becoming conceivable in the late 1860s, most promising was a system involving *stationary* steam engines that pulled cables. This was initially tried in 1867 by Charles T. Harvey on his West Side & Yonkers Patent Railway, the first of New York's elevateds. Though Harvey was an accomplished engineer, he faced difficulties with his cable system that he could not overcome. Thus credit for its practical realization went elsewhere, to an inventor on the west coast 6 years later.

Andrew Hallidie (1836–1900) was born in London and emigrated to the United States as a boy, arriving in San Francisco in 1852. After that, he spent some time prospecting in the Sierra Nevada, where he became familiar with mechanical devices used in conjunction with mining operations. Hallidie's father held English patents for techniques of spinning "wire rope," and in 1867 Hallidie himself obtained a U.S. patent for a "traveling ropeway or tramway" for transporting buckets of ore "over mountainous and difficult regions" (quoted in Kahn 1940, 28). Wire rope had previously been used in the suspension bridges for which John Roebling had become famous— work on "the Great Bridge" linking Manhattan and Brooklyn would begin in 1869. For bearing extreme longitudinal stress, crucible steels with a high tensile strength were desirable, but such steels were brittle and tended to fracture when bent repeatedly, as was bound to happen with a "traveling ropeway or tramway." Hallidie understood that strength needed to be traded off in order to gain flexibility, but he also realized that this would not entail significant danger even when the stress was greatest.

In San Francisco, there were hills so steep that an omnibus might require a 3-horse "helper" team in addition to its regular team. The steepest hills could not be negotiated by horse-drawn conveyances under any circumstances at all. But Hallidie could see how moving cables might be used on such hills—not overhead as in the Sierra Nevada, but in conduits just below the pavement, guided by pulleys. In order to impel his car forward the operator would manipulate a "grip" that reached through a slot into the conduit and clamped on to the moving cable—but clamped gradually, so as not to start with a jerk; that had been one of Harvey's problems in New York. The car would come to a stop when the operator released the grip on the cable and the brakes were applied (Figure 1.3).

With three partners, Hallidie put his first cable railway in operation in September 1873 on Clay Street, which had a 1:6 grade (that is, gaining nearly 15 feet in elevation for each 100 feet forward) as it topped Nob Hill at Jones Street (Figure 1.4). Hallidie had been anticipating a 12 percent return on his $85,000 investment, but patronage was so good that earnings topped

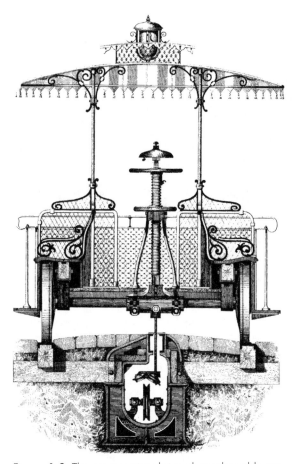

Figure 1.3: This cross-sectional view shows the cable-car grip mechanism (a wheel here but usually a lever) and the cable conduit beneath the street with only the slot open to admit the shank of the grip. (From *Street Railway Journal*, December 1884)

30 percent. Other investors rushed to climb on the bandwagon. Within a decade, five more lines had been completed in San Francisco, including a line on California Street backed by railroad magnate Leland Stanford, and several more were under construction elsewhere. Wandering around San Francisco while awaiting passage to India, Rudyard Kipling marveled at how "a cable car without any visible means of support slid stealthily up behind me and nearly struck me in the back."

Figure 1.4: The Clay Street Hill Railroad in operation. In manipulating the grip, man became so integral with machine that he himself became known as "the grip." His station was in the open car, with passengers beside him. Those who preferred to be inside sat in the trailer. Later, the two were merged in a distinctive design whose descendants ply San Francisco's cable railways to this day. (From *The System of Wire-Cable Railways for Cities and Towns*; San Francisco, 1887)

CABLE CAR TECHNOLOGY

To the mechanically minded, having a stationary power plant capable of impelling every streetcar from one end of a line to the other was indeed a satisfying concept, far more satisfying than animate power with all its shortcomings. And, compared to any other possible alternative in the early 1880s—there were several, all of them experimental—cables were the most efficient way of transmitting power for streetcars over long distances. But the ideal of "efficiency" is always a moving target, and the history of cable railways demonstrates this perfectly. What once seemed efficient and supremely rational compared to horse power would eventually seem dreadfully burdensome compared to electricity—far more readily transmitted from place to place than mechanical power could be transmitted through cables.

After San Francisco, cable propulsion was adopted in cities equally hilly, such as Seattle, and also in cities where the terrain was almost perfectly flat, such as Chicago. Chicago was actually the second city, after San Francisco, and it was quite unlike San Francisco in another way, its subzero temperatures in wintertime. Ice could build up in a conduit and inter-fere with the pulleys, and there was also a potential risk of the slot closing up as the ground froze. Engineers understood that this risk could be avoided by using extra-sturdy ironwork—very expensive, but worth it wherever rich profits could be forecast. This was certainly the case in Chicago, the fastest growing city in the country. Cable cars superseded the old Parmelee horse-car operation on State Street between Madison and Twenty-first Street in January 1882. By 1887 cable cars were hauling more than 70,000 passengers every day, and there would be as many as 100,000 daily during the World's Columbian Exposition a few years later. In his *Treatise Upon Cable or Rope Traction* a British civil engineer named J. Bucknall Smith reported that

> . . . the daily toll of over 1000 horses and 200 stablemen, etc., have been dis-pensed with, besides the continual expenses of shoeing, harnessing, sickness, food, etc., consequent upon the employment of horses. The traffic is very heavy indeed, and efficiently supports a two to three-minute service, com-monly consisting of a train comprising a dummy and three or four ordinary cars (Smith 1887, 83).

Although trains of three or four cars were eventually deemed a hazard to public safety by municipal officials, operations on the Chicago City Railway demonstrated that anywhere, in any climate, cable cars could outperform horsecars. By the time there were two more companies in operation in Chicago—the West Chicago Street Railroad running on Blue Island, Halsted, and Madison, and the North Chicago Street Rail-road running on Wells, Clark, and several other main thoroughfares—there was even a cable railway in Butte, Montana. Nobody could doubt the advantages. Cable cars could make twice the speed of a horsecar (or much more—in 1892 the *Street Railway Journal* reported that part of the Cleveland City Cable Railway's Superior Avenue line was being run at 16 miles per hour) at one-half or sometimes one-fourth the operating ex-pense. They were clean and quiet, the very soul of modernity. J. Bucknall Smith listed eighteen "points or features that may be reasonably alleged in favour of the cable system of traction" and only three "alleged objections or defects."

Yet, to repeat, the adoption of new technologies inevitably entails the acceptance of "objections or defects," another way of describing tradeoffs. Tradeoffs often involve safety, and with cable cars one of the risks (a risk not included among Bucknall Smith's "objections or defects") was the possibility of a grip getting snagged by a frayed cable strand; when a gripman

Figure 1.5: In this engraving from a contemporary magazine, the Metropolitan Street Railway's Broadway line is seen here while under construction in 1893 near Union Square. The workforce often exceeded a thousand men. When the Metropolitan switched to electricity in 1901, some economies were realized because the cable conduits were adapted to the underground wires that carried the current. Overhead wires were not permitted on Manhattan. (*Harper's Weekly*)

was unable to release the cable or stop, the scenes of panic aboard the car and then panic in the streets can scarcely be overdrawn. In financial terms, the biggest tradeoff was the initial investment, the expense entailed in digging a deep trench and installing a succession of iron yokes—hundreds or often thousands of them—to support the track, the conduit, and its elaborate conglomeration of pulleys (each of which required hand greasing through access hatches in the street). To secure them in place, the yokes would be encased in concrete. Cable railway construction on State Street required 5,000 tons of iron and steel; 50,000 wagonloads of stone, sand, and gravel; 44,000 barrels of cement; and 300,000 feet of timber to line the conduits. On Manhattan, the cost was said to be a million dollars per mile (Figure 1.5).

Then, there were the cables themselves. The Roebling firm in Trenton, New Jersey, was one of the primary suppliers. Typically, Roebling delivered cables that were slightly more than an inch thick and consisted of 96 steel wires wound into 6 strands of 16 wires each and wrapped around a hemp rope impregnated with tar, which served as a lubricant. They might stretch by as much as 2 percent as they broke in—thereby requiring continual readjustment of the tension—"but like much else about a cable . . . stretching could not be predicted with perfect accuracy" (Hilton 1982, 81). A cable *might* last for a year under ideal conditions, but one of Chicago's State Street cables got so much wear and tear that it was good for only a few months. Four miles was a typical length, Roebling's price for such a cable would be about $7,000, and the weight 60,000 pounds. There were railroad cars designed to carry that sort of load, but trucking a cable through city streets from a siding to a powerhouse could be a real challenge.

The prodigious weight, 30 tons for a 4-mile cable, is indicative of an inherent defect in the whole concept. The engine in the powerhouse had to be run at full capacity just to keep the cables moving, no matter how many cars were out on the line. The more cars actually in service, the better the economics, but the problem was inevitably more serious if a line went around corners. One expert was quoted as saying that "Curves are the bane of cable construction. Their first cost is enormous; they consume power, materially shorten the life of the rope, and are a source of endless care and anxiety to the management." The Clay Street line in San Francisco was absolutely straight, a little over a mile from Kearny to Van Ness. So were the Broadway, Lexington, Columbus, and Amsterdam Avenue lines in New York, and the long lines operated by the Chicago City Railway on State and Wabash. But Cincinnati's Vine Street Cable Railway twisted right and left, as did its Mt. Adams & Eden Park Railway, and so did the Valley City Street & Cable Railway in Grand Rapids, Michigan.

In Chicago, cable railways remained in operation for almost a quarter of a century, and the $25 million investment was well rewarded. The Grand Rapids line folded just 2 years after it opened in 1889, but other lines in small towns flourished in conjunction with the sale of subdivisions for residential housing.

The Temple Street Cable Railway in Los Angeles, for example, enabled the development of a neighborhood called Angeleno Heights where the hills were too steep for horsecars. Promotional circulars read so:

Bear in mind that this property is in the *Hills*
And on the line of the *Cable Railway System*
Have a house in the Hills!
Stop paying rent in the valleys!

Figure 1.6: Frederick Wood and John Fowler, superintendent and general manager of the Temple Street line, show off a cross-section of track and conduit, much more substantial (and costly) than the original Clay Street line. In 1892, the same year the photo was taken, Wood (at left) published a technical treatise titled *On the Temple Street Cable Railway* in Los Angeles, California, with illustrations of numerous refinements that he and Fowler had devised, including a double-jaw side grip that obviated the need for a turntable at each end of the line, as in San Francisco. Alas, no matter how ingenious these two men may have been about mechanical matters, cable propulsion was on the verge of being rendered obsolete by electricity. (Security First National Bank photo, author's collection)

This cable railway remained in operation for 16 years, from 1886 to 1902, during which time the population of Los Angeles more than doubled and Angeleno Heights became a prime residential address. The cost of constructing the line was a write-off, but, for investors, that was incidental; the profits were in the residential homesites that could not have been developed otherwise (Figure 1.6).

Figure 1.7: A timeless sight: the grip swings his car around on one of San Francisco's turntables, this one at Hyde and Beach streets, at the foot of the 20.67 percent Hyde Street descent of Russian Hill—the most spectacular of all views of San Francisco Bay and the inspiration for the "Ballad of the Hyde Street Grip." Built as part of the California Street Cable Railway in the 1880s, this line is one of the three now operated by the San Francisco Municipal Railway and considered to be one of the city's premier tourist attractions. (Author's photo)

Ultimately there was at least one cable car line in every large American city except Boston, Detroit, and New Orleans. In several cities they played a role similar to what they played on Temple Street, in others they were profitable ventures in and of themselves, as in Chicago. Cable cars crossed the Brooklyn Bridge until 1908, a line in Seattle lasted until 1940, and of course three of San Francisco's lines remain in operation to this day (Figure 1.7). But new construction had all but ceased within 20 years after the Clay Street line opened. The one major exception was Manhattan, where there were delays because of patent litigation, and as a result these very expensive lines remained in operation for only a few years before getting converted to electricity. At the peak of cable railway operation nationwide in 1893, there were 305 miles of double-track line, but that number would fall to 101 some 10 years later and would be only 20 in 1913 (see Table 1.1). In 1913 that was only a fraction of the remaining horsecar mileage, and there were trolley-car systems with more than 300 miles of line in several American cities. Even as Andrew Hallidie himself had heralded cable railways as "the true solution" to the problem of urban mass transit (Hallidie 1885, 13–14), other men had kept searching for a mode of propulsion that did not entail

Table 1.1
Cable railway mileage, 1873–1913 (double track)

	Miles built	In service	Out of service
1873	0.6	0.6	
1877	1.5	2.1	
1878	2.5	4.6	
1879	2.4	7.0	
1880	4.2	11.2	
1881	...	11.2	
1882	9.1	20.3	
1883	10.8	31.1	
1884	...	31.1	
1885	15.4	46.5	
1886	15.8	62.3	
1887	32.2	93.0	1.5
1888	59.3	151.4	0.9
1889	66.1	213.0	3.7
1890	54.3	266.4	1.7
1891	21.9	272.3	16.0
1892	22.4	287.6	7.1
1893	38.2	305.1	10.7
1894	6.8	302.3	9.6
1895	7.5	289.5	20.3
1896	...	255.7	33.8
1897	...	241.7	14.8
1898	...	220.2	21.5
1899	...	183.8	36.4
1900	...	147.9	35.9
1901	...	120.3	27.6
1902	0.4	108.4	12.3
1903	...	101.4	7.0
1904	...	95.3	6.1
1905	...	95.3	...
1906		29.9	66.0
1907	...	29.3	...
1908	...	28.2	1.1
1909	...	27.7	0.5
1910	...	25.9	1.8
1911	...	25.0	0.9
1912	...	21.0	4.0
1913	...	20.0	1.3

Source: George Hilton, *The Cable Car in America* (1982).

such drastic tradeoffs—that was faster, cheaper, and less cumbersome. But any "true solution" had to match a cable railway's reliability, and for a time that proved elusive. It was coming closer to reality even in 1885, however, and within only a few years such a mode of propulsion had been unmistakably identified and proven.

2

The Trolley Ascendant

A handy framework for analyzing technological change was devised by the foremost historian of aeronautical engineering, Walter Vincenti: technologies have an early stage, he writes, when "the knowledge sought is that of a workable general configuration," and one or more subsequent stages, when inventors are seeking "a particular instance" of that configuration (Vincenti 1990, 241–250). The reference is to airplanes, and to the transition from gasoline engines with propellers to turboprops and then to jets, but the framework is equally applicable to streetcars. The workable general configuration appeared between the 1830s and the 1850s, with iron rails and flanged wheels, and a lightweight design emerged that was distinct from the design of railroad cars but still used horse power. Many other "particular instances"—that is, other possible ways of impelling streetcars—were being explored. One of them, underground cables, was successful, at least for a time and under certain circumstances: on Clay Street in San Francisco, State Street in Chicago, and Temple Street in Los Angeles, for example, if not on the Valley City Street & Cable Railway in Grand Rapids.

But other instances strike us today as obviously impractical. "No horses to buy and feed, no cable to renew, no stables to rent or stablemen to pay. No steam to frighten horses, no disagreeable dust, no fire and no smoke. No electric shocks to frighten ladies and endanger life." So read an 1885 prospectus for a streetcar impelled by compressed air. A prototype was actually constructed, as was an experimental streetcar with ammonia gas as "an active propelling agent" and another using a volatile petroleum distillate called naptha (Rowsome 1956, 35–48). Impractical, yes, and probably even more impractical was a streetcar outfitted with a very large spring wound taut by a stationary engine (which never got an actual tryout). But the history of technology is full of evidence that *invention* is easy compared to the process of rendering an invention practical in economic terms. Or even easier than conceiving that it *might have a practical use*. The reference to "electric shocks" in the compressed-air prospectus is suggestive.

ELECTRICAL PIONEERS

In 1799 Alessandro Volta (1745–1827) discovered that electric current was "excited by the mere contact of conducting substances of different kinds," what came to be called electrodes and electrolytes. This was the basis of the electric battery. Twenty years later, Hans Christian Oersted (1777–1851) discovered that a magnetic needle was deflected at right angles to a wire carrying an electric current, and in 1820 André Marie Ampére (1775–1836) described the "mutual interaction" between current-carrying wires. This was the basis of the science of electromagnetism. In 1831, Michael Faraday (1791–1867) demonstrated that a current could be generated in a wire by sending current through an adjacent wire or by moving a magnet in the vicinity. This was called electromagnetic induction. These were all important contributions to the store of scientific knowledge, but so far as we know none of these men—who were known as *natural philosophers*—ever thought about practical applications. That fell to a younger generation of men who would become known as *electricians*. Could the current from a battery be combined, for instance, with the force of electromagnetic induction in order to derive "an active propelling agent?" In the expression of the time, was it possible "to harness the forces of nature?"

In 1842, a Scotsman named Robert Davidson (1804–1894) built a small battery-powered electromagnetic locomotive and received permission to take it out for a test run on the Edinburgh & Glasgow Railway. When he could get it to go no faster than 4 miles per hour, a disappointed observer remarked that its performance was "rather unfavorable to the claims of electricity" (quoted in Post 1974, 13). But 9 years later, in 1851, an American electrician named Charles Grafton Page (1812–1868) demonstrated a much more elaborate and powerful electromagnetic locomotive that managed an 11-mile round trip from Washington, D.C., to Bladensburg, Maryland. "Propelled by some invisible giant, which by his silence was as impressive

as his noisy predecessor," it reached speeds as high as 19 miles per hour
(quoted in Post 1976, 98). Page knew as much about electrical science and
technology as anyone in America, and his locomotive clearly worked in a
"technical" sense. But it did not work well enough to catch the eye of anyone eager to invest in its improvement. (To this point, Page's experiments
had largely been financed by the federal government—a precedent for federal sponsorship of R&D for conveyances that were unattractive to private
investors: the Standard Light Rail Vehicle [SLRV] from Boeing Vertol in
the 1970s or the Advanced Technology Transit Bus [ATTB] from Northrop
Grumman of the 1990s, for example.) Page's was a signal demonstration that
electromagnetism had the capacity to impel a vehicle, but more than anything else he demonstrated something not intended, that the tradeoffs were
simply too extreme: As a source of electrical current, batteries had overwhelming limitations, notably the rapid depletion of the electrodes (the
electrolytic action literally ate them away) and the excessive weight.

In the early years of the twentieth century, streetcars using a different type of batteries called storage batteries (the kind with which we are
presently familiar in automobiles) would prove feasible and would be operated on a minor scale on the streets of lower Manhattan, where the traffic was
too light to warrant the expense of electrification. Because overhead trolley
wires were considered unsightly, battery cars were also seen around residential tracts and resorts during the boom in Florida real estate in the 1920s.
But in the latter nineteenth century there was a widely shared assumption
about conveyances powered by electricity: If they were ever to "work" in an
economic sense as well as a technical sense, they would need to be designed
to derive their current from a stationary generator that was turned by a stationary steam engine, of the sort used with cable car lines. Such generators,
called dynamos, were not rendered practical until the 1870s, and still there
was a big challenge: how to transmit current from such a stationary source
to the motor in a *moving* conveyance—and then back where it came from
in order to complete the circuit. As an early treatise on the technology of
electric railways put it: "an *uninterrupted metallic circuit* must extend to the
car and back from the car to the dynamo" (Whipple 1889, 71).

MAKING A CIRCUIT

In the summer of 1879, an International Trades Exposition was staged in
Berlin, Germany. Among the featured attractions was an electric railway
designed by Ernst Werner von Siemens (1816–1892), whose name is still
associated today with electrical technologies, not least with light railway

vehicles (in *Jane's Urban Transport Systems*, Siemens Transportation Inc. in Sacramento, California, advertises "turnkey LRV systems"). Visitors to the exposition could actually ride in diminutive cars behind a diminutive electric locomotive: It was a toy, but it was also much more than a toy, because von Siemens had gone a long way toward solving the problem of making "an uninterrupted metallic circuit." He sent 150 volts from his dynamo to the locomotive via a conductor between the rails, and sent it back—that is, grounded it—via the rails themselves. Others followed suit in the United States. In 1880, Stephen D. Field (a nephew of Cyrus Field, who financed the Atlantic Cable) built an electric railway at Stockbridge, Massachusetts. And none other than Thomas Edison built one at Menlo Park. Both had some distant promise of practicality, and Field and Edison later pooled their resources in a company backed by Henry Villard of the Northern Pacific Railroad—who envisioned, as had Charles Grafton Page, electrified railways in the arid West where fuel and water for steam locomotives were in short supply.

But nothing came of these efforts, nothing that worked in any more than a "technical" sense. A major practical concern was safety, the least palatable of all tradeoffs, as the history of the steam locomotive in urban transit demonstrates. Voltage is the electrical equivalent of pressure, and completing a circuit over any appreciable distance requires high voltage, much more than the 150 volts that von Siemens used in Berlin. Any high-voltage circuit involving the running rails and a third rail between them (or alongside) posed a danger to any living thing that might come in contact with both halves of the circuit at once—and to this day the expression "third rail" still implies grave and immediate danger. When elevated rapid transit lines were electrified beginning in the 1890s, they typically had 600-volt third rails, but the high platforms in stations kept people from getting in harm's way. Anywhere that the running rails were to be at street level, it was essential that the "live" half of the circuit be kept out of reach. There were two possibilities, either in a covered conduit below the surface or else overhead.

Although neither von Siemens nor Field and Edison were in a position to expend further time and energy on this problem, other electricians jumped at the chance. And even though their names are less familiar today, their efforts are nonetheless historic. One was Charles Van Depoele (1846–1892), a Belgian who had emigrated to the United States in 1869, prospered in furniture manufacture, and then gone into the fledgling arc-lighting business in Chicago; another had arrived in America from Birmingham, England, a few years before Van Depoele, and he had an unusual surname, Daft, a word the dictionary defines as "simple or foolish." But Leo Daft (1843–1922) was anything but daft in that sense; indeed, he would come close to achieving lasting fame as the inventor of the electric streetcar.

While Daft and Van Depoele were the two names most in the news during the mid-1880s, they were not the first to attempt to employ electricity in commercial urban transit. In Germany, von Siemens tried to do it, and in Cleveland, Ohio, so did Walter Knight and Edward Bentley, who developed their equipment in the shop of Charles Brush, a manufacturer of arc-lighting dynamos. In 1884, Knight and Bentley arranged to take over a 2-mile segment of the East Cleveland Horse Rail Company. They transmitted power via an underground conduit, and contact was made by means of a "plow" that protruded underneath the car, just like the shank of a cable car grip. This setup, visually indistinguishable from a cable railway, would eventually be brought into the realm of practicality, to be employed in cities where overhead wires were perceived as a nuisance. But Knight and Bentley did not have adequate financing to keep their Cleveland system operating in the face of many technical difficulties. Very likely they learned important lessons that they applied when they built a similar conduit operation on the Observatory Hill Passenger Railway in Allegheny City, Pennsylvania. Yet they failed to keep that railway operating either.

Solving problems connected with "electromagnetism as a motive power"—safety, reliability—would entail confronting many challenges. Van Depoele and Daft both had better financial resources than Knight and Bentley, and both had already staged successful demonstrations of electric locomotives that picked up power from a third rail beside the tracks, Daft on the Saratoga & Mt. McGregor Railroad in New York, Van Depoele at an industrial exposition in Chicago, where he had a manufacturing plant. They envisioned these locomotives being used on elevated railways, but they were both thinking about street railways as well. In 1885, Van Depoele began a commercial operation on the streets of South Bend, Indiana, that drew current from overhead wires. Within a short time, he also had electric streetcars operating in Scranton, Pennsylvania, and Appleton, Wisconsin. These lines were frequently shut down for repairs, however, whereas Daft had managed to keep an electric railway in Baltimore in continuous operation for more than a year.

LEO DAFT AND THE BALTIMORE UNION PASSENGER RAILWAY

In addition to his knowledge of electrical technology, Daft had the practical sensibility of a successful businessman, having operated an electrical manufacturing and supply business in Greenville, New Jersey, since 1881. The New York Power Company was a regular customer, and in 1884 Daft's firm had provided the equipment for a complete central generating station for the Massachusetts Electric Power Company. These were electric lighting companies, but the challenges of generation and distribution were somewhat similar to electric transit. In 1885 Daft signed a contract to electrify a horsecar line—actually, it had been using mules, like Booth Tarkington's Midland—the Hampden branch of the Union Passenger Railway in Baltimore. To generate electricity, he connected a stationary steam engine to two dynamos supplying 260 volts. (The connection was provided by means of heavy leather belts, which were commonly used in factories to link machines by means of shafts and pulleys to central steam engines.) To provide motive power out on the line, he built a pair of boxy "tractors" that were outfitted with 8-horsepower motors; the plan was to couple these tractors to cars formerly drawn by mules. To transmit power to his tractors he used a third rail. Although he knew that third rails were problematic with street railways, he assumed, as had von Siemens, that the relatively low voltage would minimize danger. The line was opened for business on August 15, 1885.

According to Daft's contract, full payment was contingent on a year of successful operation, and he did manage to keep the line operational. But "success," like "efficiency," is a word with an infinitely flexible meaning, and Daft achieved success only because he was prepared to pay constant attention to the temperamental motors. If motors ran at a constant speed, as in a factory, their performance was satisfactory. But frequent starting and stopping caused all sorts of problems with the brushes, which made contact between the static part of the motor and the rotating armature. Even more importantly, Daft realized that he had miscalculated the potential hazards of using a third rail, even with low voltage, and that he was going to have to use overhead wires instead.

In late 1886, when Daft built an electric railway on Pico Street and Maple Avenue in Los Angeles, and another at Sea Girt, New Jersey, the entire electrical circuit, both the live wire and the ground, was overhead. It was carried through a pair of wires on top of which was a 4-wheeled rolling device at the end of a flexible cable. Placing a pair of wires far out of reach took care of most concerns about safety, and also got around a serious problem when the rails were used to ground the circuit: unless there was a tight bond between the joints, current would jump to anything metallic such as water mains or gas pipes. So, in Los Angeles and Sea Girt, Daft would have no problems of that sort, but his system was still unreliable. The trollers had a tendency to come off the wires. Moreover, he still had not solved the problems with the motors—their tendency to short circuit—and with the way that they were connected to the axles and wheels. Might he have been able to surmount all these problems? The history of technology is full of "might haves." Ultimately he faced the same impediments that Knight and Bentley, Van Depoele, and others had failed to conquer—impediments that were as much financial as they were technical. By 1888, even as his line in Los Angeles was foundering after 18 months of off-and-on operation, another firm had come along whose principals had both the expertise and the capital to assure the success of electric traction.

FRANK SPRAGUE AND THE RICHMOND UNION PASSENGER RAILWAY

Leo Daft's Hamden line in Baltimore was "the first regularly operated electric road in this country"—these are the words of an inventor named Frank Sprague who was Daft's rival in the early days of electric traction (Sprague 1905, 444). And yet it is Sprague's name, not Daft's, that is most enduringly linked with the invention of the electric streetcar. Sprague (1857–1934) was a graduate of the U.S. Naval Academy in Annapolis. After leaving active service as an engineering officer, he worked for Edison in the early 1880s setting up generating plants for electric lighting systems. Then he moved on to form his own company, the Sprague Electric Railway & Motor Company. In May 1887 he electrified a single-track streetcar line in St. Joseph, Missouri. Then, he signed a contract with a group of New York investors for a much larger and more difficult project, to electrify another Union Passenger Railway—this one in Richmond, Virginia—that was in the process of laying 12 miles of double track throughout the city. Richmond was to be a real transit *system*, not just one isolated line, as in St. Joseph or with most of Daft's and Van Depoele's projects. Sprague was to design and equip a generating station and rig a full complement of overhead wires. He was to outfit each of forty cars with two motors and all appurtenances necessary for their operation. If everything was satisfactory, Sprague would receive $110,000.

SPRAGUE'S TECHNOLOGY

It was an arrangement, Sprague later remarked, "that a prudent business man would not ordinarily assume." But Sprague was something of an adventurer, and so were the two men he hired to assist him, one a fellow graduate of the Naval Academy, the other from West Point, both with "energy, pluck, and endurance." Sprague was already manufacturing 500-volt motors for factory operation that he thought would do the job in streetcars. Each car would have two axles, each with its own motor. One end would be connected to

Figure 2.1: Frank Sprague's technique for mounting motors became conventional with all streetcars. As a contemporary technical treatise put it, they did not "take up any sitting or standing room" in the car itself and they ensured "perfect parallelism in the meshing of the gears." The weight of each motor was distributed between the bar in the center of the frame and bearings wrapped around the axle. (Frank J. Sprague, "The Story of the Trolley Car," *The Century Magazine*, August 1905)

the frame that held everything together, the other end would be supported on the axle itself. This would enable the gears to remain engaged even as the wheels moved up and down when they rolled over irregularities in the tracks. Before Sprague, motors had often been mounted inside the car or on the front platform, for ease of access when the brushes needed attention, as they so frequently did. They would be connected to an axle with transmission belts or with chains on the order of bicycle chains, which were subject to frequent failure. Direct gearing to the axles would be a far more trustworthy arrangement (Figure 2.1).

Another feature of Sprague's design was the arrangement for picking up power via a pole atop each car, at the end of which was a little wheel that was *pushed up under an overhead wire* by means of a spring at the base of the pole. Not very many inventions can be attributed to "a flash of insight," even though we hear this expression all the time; indeed, the electric streetcar is a fine example of invention by incremental improvement, not sudden insight. Without the spring-loaded trolley pole and under-running trolley wheel, there would never have been street railways in every American city and town. It was a remarkable idea. The interesting thing, however, is that it was not Sprague's idea originally, but rather Van Depoele's. In the course of patent litigation, Van Depoele testified that in 1883 he had devised "a contact roller ... mounted on a beam which was pivoted at its center and provided at one end with a spring pressing the contact up

Figure 2.2: In this engraving showing a Daft operation in a town on the New Jersey sea-coast, a troller can be seen above the car on the left, towed along by flexible cables but nearly always in danger of falling off. The folklore of pioneer Daft and Van Depoele lines included stories of boys with rubber gloves (the two wires made a "live" circuit) who were paid to ride rooftops and put errant trollers back where they belonged. (Fred H. Whipple, *The Electric Railway*, 1889, reprint 1980)

against the underside of the wire." But Van Depoele did not realize what a good idea he had. Instead, his cars, like Daft's, towed a 4-wheeled device with grooved wheels along paired wires: a "roller" or a "traveler." Or a "troller."

This arrangement had one definite advantage—because both the positive current and the return were overhead, there was no need for establishing a ground through the rails, which, to be truly effective, required bonding each length of rail to the next one with a copper cable. But, as so often, there was a significant and eventually decisive tradeoff—the tendency of the troller to get dislodged from the wires and come crashing down on the roof of the car (trollers were heavy, the resemblance to a roller skate only superficial) (Figure 2.2). In contrast, the under-running wheel rarely came off the wire, and, if it did, it could be put back simply by pulling on the end of the pole with a rope. It was the same as a troller in one respect: It was dragged behind, in the manner of a fisherman trolling. Eventually, though nobody has ever explained exactly how, the word would become "trolley" and it would define the entire conveyance. A trolley was an electric streetcar.

Electric streetcars were not at all uncommon by 1887 when Sprague started to work in Richmond—besides lines in Wisconsin, Pennsylvania, California, and New Jersey, one could find them in Detroit and Port Huron, Michigan; Windsor, Ontario; and Montgomery, Alabama. But they were everywhere notorious for their unreliability. Motors would short-circuit, trollers would topple off the wires, chain drives would fail and fling broken ends around dangerously. Sprague's motors were relatively well engineered and there were no trollers and no chains. He had reason to hope for the best when he opened his system to the public in early 1888. But there were many unwanted results at first. Cars would stop suddenly as the gears froze up. Sprague was afraid that they had not been properly machined, but his mechanic Pat O'Shaughnessy understood that the difficulty was nothing more serious than insufficient lubrication. There were other problems with the commutators, the device that translated direct current into alternating current. To get proper tension, Sprague had to rework the spring-loading mechanism at the base of the trolley pole over and over. Funds dwindled away, but they never ran out, and by May 1888 he had all forty cars in service.

As suggested by his remark about Daft's Hamden line in Baltimore, Sprague himself understood that he did not merit sole credit for the invention of the electric streetcar. Indeed, this remark appeared in a long article in which he honored the efforts of electricians as far back as Robert Davidson and Charles Grafton Page. But it is truism that recorded history typically sweeps together the accomplishments of many inventors in order to denominate one particular individual as *the* inventor of a particular device: telegraph, Morse; light bulb, Edison; telephone, Bell. Frank Sprague had persevered, and it was Sprague who would "go down in history" like Morse, Edison, and Bell. What had really happened was a more complex story—more "messy," as historians of technology like to say—and yet it was clear to Sprague's contemporaries that he had accomplished something that others had not. Most importantly, it was clear to Henry Whitney, who had consolidated a number of Boston horsecar lines into the 212-mile West End Railroad, "the largest single street railway system in the world and the first to provide unified operation of public transit in a major American city" (Cheape 1980, 116). In the summer of 1888, Whitney, along with his general manager, Daniel F. Longstreet, paid a visit to Sprague in Richmond. Longstreet had recently resigned as general manager of the horsecar company in Providence after a rival firm secured a franchise for a cable railway. He did not think that old New England cities, with their crooked and narrow streets, were appropriate for cable car lines. But Longstreet was also skeptical about electrification. What would happen, he asked Sprague, if the tracks got blocked and a long line of streetcars backed up in one place?

Dealing with that contingency had been part of Sprague's contract, so he already knew what would happen. One evening he gathered together all the motormen and had them take 22 cars out on the line and stop them one behind another. And then he showed Longstreet and Whitney that they could all be started at once. Whitney, a wealthy man who had inherited the presidency of a steamship company, was a speculator in Boston real estate stretching westward along Beacon Street into the town of Brookline. He had been developing plans for running cable cars from Adams Square to Egleston Square in Roxbury, and also from Bowdoin Square across the Charles River to Harvard Square in Cambridge. After visiting Richmond, he shelved these plans. Sprague's company soon had a contract to provide the electric motors that would eventually displace 8,000 horses on the West End Railroad. Soon after that, he had more than a hundred other contracts, and he got an offer he could not refuse, to merge his company with Edison General Electric.

Later, Sprague would turn his attention to Chicago's Metropolitan West Side Elevated Railway and develop the so-called multiple unit system, as important an innovation as his trolley system in Richmond. The two had one thing in common—they both enabled conveyances to apply "traction" through their own wheels, rather than deriving it from an external source, as with cable cars. During the early 1890s, street railway executives in San Francisco, New York, and Chicago, all heavily invested in cable railways, continued to argue that they were superior to electric streetcars wherever there was exceptionally heavy patronage. And it was true that a cable railway could almost never be overloaded; on its State Street line, the Chicago City Railway operated hundreds of trains (grip car and trailers) simultaneously during rush hours. But their arguments became less and less germane as more powerful electric motors became available from General Electric and Westinghouse, and GE introduced what it called its Type-K motorman's controller to regulate speed, a device that was so durable and dependable that it would be used on nearly all streetcars until the 1930s.

OVERHEAD WIRES OR CONDUITS?

The electrified Chicago "L" had a third rail at trackside, but that was permissible only in places where cars would be boarded from high-level platforms and people could not inadvertently come in contact with the rails. The Richmond system was like any other street railway—the cars ran in the street—and it was essential to protect people from an accidental encounter with several hundred volts of electricity. In Richmond there were trolley wires 14 feet overhead, attached by means of hangers with an inverted u-shape to cables spanning the street from poles on either side. Current was distributed by means of an insulated feeder strung from one pole to the next all over the system, and current was returned through the rails, which were spiked to wooden crossties and bonded from the end of one section to the next in order to provide a continuous circuit.

Sprague understood that concerns about the safety of overhead wires, and also about their appearance, had not been put to rest. He was therefore uncertain whether this arrangement would be acceptable everywhere and it turned out that it was not. Overhead wires were prohibited by municipal authorities in London, Paris, Marseilles, Berlin, Budapest, and Prague, and also on Manhattan and in Washington, D.C. If there were to be electric street railways there, current would have to be distributed via underground conduits—expensive and complicated, the last thing in the world any cost-conscious transit manager or practically-minded engineer would have chosen when there was the comparatively simple option of trolley wires. But that option was not always open. Eventually there were more than 200 miles of streetcar line on Manhattan with conduits between the rails. Because of the cost, some of the cross-town horsecar lines were not converted until well into the twentieth century, and others were simply abandoned with no replacement at all. In Washington, D.C., there was less mileage with conduits, only about half as much, because a conventional overhead trolley wire was permitted in outlying areas—on Wisconsin Avenue beyond Georgetown, for one. Where there was an interface between underground and overhead electrical distribution systems, workers were stationed in pits under the tracks to affix and detach conduit plows while the conductor raised or lowered the trolley pole.

The engineer responsible for the Washington system, Albert Connett, was familiar with the concept because he had been involved in Bentley and Knight's short-lived conduit operation in Cleveland in 1884. Later, he supervised the construction of conduit lines in London and other cities overseas. No doubt the vista was less cluttered in cities that prohibited overhead wires, but it could be argued that rational technological choice had been traded away for considerations that were "merely" aesthetic and therefore trivial; the relative costs of construction was one of the major points of comparison between cable railways which required conduits and trolley lines which did not. True, the maintenance of trolley wires was always fussy, especially at intersections and curves, which required a spidery network of supporting cables. Tension had to be adjusted and readjusted, and

Figure 2.3: Seen here in the 1940s is the intersection of 14th and G Streets NW in Washington, D.C., where two lines crossed and there was a set of connecting tracks. Laborers are making repairs both above and below the surface. While the design of the electric conduit system is not fully evident in this view, what is visible suggests an astonishing complexity. Note, however, that there is not one wire to be seen anywhere above ground. (Capital Transit Company photo, author's collection)

of course trolley wires would eventually wear out and need replacement. But dealing with even the most complex situation—at a "grand union" where two double-track trolley lines crossed each other and there were connecting tracks in all four directions—was easy compared to what conduits entailed. The demands of routine maintenance were relentless. Jammed plows were an everyday occurrence, and renewing trackwork was a truly formidable undertaking (Figure 2.3).

An unusual technological choice of another sort took place in the handful of cities that had neither conduits nor a conventional single trolley wire. Rather, there were paired wires. Later, paired wires would be essential with trackless trolleys because there were no rails to conduct the return current. But with streetcars? The biggest city to have such a system was Cincinnati, Ohio. When horsecar lines were slated for electrification, as was the case in Cincinnati in 1889, the utility companies would often express concerns about return current in the rails "wandering away to follow paths other than the track" (Miller 1943, 138). When trolley cars were still a novelty it was not difficult to raise popular alarms. In Cincinnati, using the tracks for return current was first banned by an injunction and then forbidden by municipal ordinance. Or, rather, tearing up the pavement to bond the rails that had initially been used by horsecars was forbidden. The second trolley wire was intended as a temporary expedient until "the war of the wires" could be resolved, as it was. But after paired wires had been installed all over town, the operating company concluded that the system might as well remain as it was, and so it did until the end of trolley operation in Cincinnati (White 2005).

Elsewhere, it became standard procedure to transmit the live current through a single overhead wire and ground it through the rails. Electrolytic damage to buried water and gas pipes never became a critical problem, and, with most streetscapes already cluttered with wires and poles for telegraphs, telephones, and electric lights, adding some more clutter did not seem to make much difference. It was a tradeoff that most people were willing to accept. Along with New York and Washington, Chicago at first prohibited trolley wires, but later relented. In those three cities the transit companies were temporarily ahead of the game because the old cable conduits could be refitted to carry electricity. But cable conduits were integral with the yokes that also supported the trackwork, and everything soon had to be replaced when the rails proved inadequate to supporting newer trolley cars, which were much heavier than cable cars that lacked any mechanical equipment except for the grips and brakes.

For a time there were reports of people getting traumatized by the electric arcs that resulted when a trolley wheel momentarily lost contact with the wire—a phenomenon that would occur right outside the windows of second-story offices and hotel rooms. But rarely were these reports credible, and in the long run safety considerations weighed heavily in favor of trolleys in comparison to cable cars and even horsecars. By the time of the 1890 census the total mileage of trolley car lines nationwide was four times that of cable railways. Even though horsecars were still prevalent in some cities and towns, there was little reason to doubt the president of the American Street Railway Association (ASRA) when he announced at the association's annual convention in 1890 that he was "thoroughly convinced that electricity is the coming power" (quoted in Hering 1892, 24).

It was for sure, and soon the ASRA had been renamed the American Electric Railway Association, AERA. A 1902 census put the mileage of electrified lines in the United States alone at nearly 22,000, and more than 90 percent of all streetcars were electric. Except on Manhattan and in a handful of small towns, horsecars had all but disappeared, and so had all but 108 miles of cable railway. Most of the remaining mileage in the city with the most lines, San Francisco (52.8 miles all told), would be wiped out by the earthquake and fire in 1906. Except for San Francisco and Seattle, the cable railways that remained in operation longest were all overseas: in Edinburgh until 1923, Paris until 1924, the Isle of Man until 1929, the comprehensive system operated by the Melbourne Tramway & Omnibus Company until 1940, and the Mornington Tramway in Dunedin, New Zealand, until 1957.

STREETCAR SUBURBS

In the quarter-century after Frank Sprague rendered the trolley car a practical mass-transit conveyance, the population of the United States grew from less than 60 million to more than 95 million. In 1890 the size of the rural population was still double the size of the urban population; by 1920 the size of the urban population had caught up with the rural population and indeed exceeded it—54,158,000 people lived in "urban places," 51,553,000 in "rural places."

By 1940 it was far ahead (see Table 2.1). Electric street railways were integral with urbanization in many parts of the world, but nowhere to such a marked degree as in the United States. This was the epoch of America's rise to world power. After victory in the Spanish–American War in 1898, the United States had acquired an empire reaching halfway around the globe, Puerto Rico, Guam, the Philippines. Steam railroad mileage stretched toward 250,000. In steel, textiles, and many other industries, the United States

Table 2.1
U.S. population, urban and rural territory, 1870–1940

	Urban	Rural
1870	9,902,000	28,656,000
1880	14,130,000	36,036,000
1890	22,106,000	40,841,000
1900	30,160,000	45,835,000
1910	41,999,000	49,973,000
1920	54,158,000	51,553,000
1930	68,955,000	53,820,000
1940	74,424,000	57,256,000

Source: Benjamin J. Wattenberg, *The Statistical History of the United States* (1976).

dominated markets worldwide. Burgeoning industries demanded larger and larger workforces. Consumer-goods factories converted their power supply from steam engines, line-shafting, and leather belts to electric motors. A single factory might employ tens of thousands of workers, and white-collar workforces grew apace. As steel-framed construction techniques were developed and "skyscrapers" began to command the vista in city after city, there might be thousands of men and women working in a single office building.

As to housing, for unskilled workers the conditions were almost beyond belief. Immigrants arrived by the millions, most of them with only a few dollars in their pocket. They had nowhere to live but tenements, "great prison-like structures of brick, with narrow doors and windows, cramped passages and steep rickety stairs ... had a foul fiend designed these great barracks they could not have been more villainously arranged to avoid any chance of ventilation" (Hall 1998, 752). It was no longer even remotely possible to crowd housing for both the working class and the white-collar class inside the bounds of the "walking city." Streetcar lines were designed with the express purpose of opening up new residential tracts whose appeal would be to people of some means who "were repelled by conditions in the central city." They were called streetcar suburbs.

In streetcar suburbs, single-family homes, duplexes, or "three decker" apartments were the norm, with retail activities such as grocery stores and pharmacies clustered around the streetcar stops. "Whether a man lived in a lower middle class quarter of three-deckers, or on a fashionable street of expensive singles, the latest styles, the freshly painted houses, the well-kept lawns, and the new schools and parks gave him a sense of confidence in the success of his society and a satisfaction at his participation in it" (Warner 1962, 156). None of this transformation could have occurred without the range, capacity, and speed of the electric streetcar. There are few better examples of a mutual interaction between technology and society.

Through restrictive covenants on the sale of homes and homesites, the new suburbs facilitated residential segregation based on race or ethnicity. They separated families on the basis of income because of the way that the price of housing varied from one neighborhood to another. There were many differences among the communities that grew as streetcar suburbs, some within the bounds of old city limits, others beyond: Evanston and Oak Park (Chicago); Silver Spring and Takoma Park (Washington, D.C.); Eagle Rock and Huntington Park (Los Angeles); Morrisania (the Bronx); Mount Lebanon (Pittsburgh); Chestnut Hill and West Philadelphia; Cleveland Heights and Shaker Heights. But there were similarities as well. "Park" was a favorite name, suggesting a place of peace and quiet. "Heights" and

"Mount" both suggested a place above it all; in Los Angeles, Angeleno Heights, a new streetcar suburb of the 1890s, was just a few miles from Mount Washington and Highland Park, which were subdivided a few years later. Roland Park, on the northern fringe of Baltimore, was developed at the same time as Angeleno Heights by Edward H. Bouton (1860–1941), a Kansan who had previously been involved in another Los Angeles subdivision called the Electric Railway Homestead Association. To lay out plats for Roland Park, Bouton enlisted Frederick Law Olmsted, the designer of Central Park in New York City and the country's most distinguished landscape architect. Bouton himself established the street railway connecting Roland Park with the Baltimore harbor-front. Later, he developed other Baltimore streetcar suburbs such as Guilford, Homewood, and Northwood, and he advised developers in the suburbs of Chicago, Cleveland, and Houston on the integral role of streetcar lines and residential subdivisions.

Perhaps the quintessential streetcar suburbs (and the subject of a classic history with that same title) were Roxbury, West Roxbury, and Dorchester, formerly separate towns that when annexed to Boston "more than doubled the area of the whole 1850 pre-streetcar metropolis" (Warner 1962, 35). Henry Whitney owned expansive tracts throughout the area, providing transit to and from Boston Common on his West End Railway. During the last three decades of the nineteenth century, thousands of residential permits were issued for these suburbs, and by 1900 they would have a population of a quarter million. Fifty years later, they would have Boston's largest residential concentration and be integral with the "central city" from which they had once been distinct. New development had long since moved further out into less crowded and now more desirable streetcar suburbs, perhaps in the other direction, to Arlington Heights, say, or Medford, which were also well served by trolleys.

As suggested by the activities of Bouton and Whitney, suburban streetcar lines and residential development were typically in the hands of the same men. The most notable of these men on the west coast was Henry Huntington, whose Land and Development Company controlled suburban tracts in Southern California's San Gabriel Valley, in the San Fernando Valley, and along the Los Angeles and Orange County beachfront, all of which could be reached by his expansive Pacific Electric (PE). Huntington regarded the PE more as a means to the end of selling land than as a moneymaker itself, and eventually sold it to the Southern Pacific Railroad, for hauling freight as well as passengers. But Huntington also owned another system, the one that provided local transit in L.A., the Los Angeles Railway, which he viewed as a profitable enterprise and kept until his death in 1927. Huntington was one of many capitalists who gained control of local transit firms that formerly had competed with one another, and also built trolley lines with the aim of driving horse or cable lines out of business. The original name of the Los Angeles Railway in 1891 had been the Los Angeles Consolidated Electric Railway, and it gained control of horsecar lines dating back to the 1870s and cable railways built in the 1880s.

The word "union" or "consolidated" in the name of a transit firm was a sure sign that it had pulled together other lines that had formerly competed with one another. Typically the plan after a consolidation was for energetic expansion and heavy investment in new equipment. In the years before World War I, the expectation—actually, the faith—was that ridership and profits, even with 5-cent fares seemingly inviolate, would continue to grow far into the future. Trolleys, much faster that horsecars or cable cars, greatly increased the distance that seemed practical for traveling to and from work, or, in the expression favored by white-collar workers, *commuting*. In the 1870s, a Los Angeles commuter, a merchant or a clerk, might come in by horsecar from Boyle Heights, a couple of miles away across the Los Angeles River; by the 1890s it seemed reasonable to consider commuting from Garvanza, halfway up the Arroyo Seco to Pasadena, or even from Pasadena itself, 9 miles away in the San Gabriel Valley.

DOUBLE TRUCKS

The growth and expansion of population into outlying areas raised demands for better service, and streetcar companies were often compelled to make improvements by the terms of their franchises. There had to be more car stops, stops that were closer together. Service had to be more frequent, cars larger. The first trolleys usually had four wheels attached to pedestals that were integral with the body, the same as most horsecars; indeed, some of the earliest electric streetcars were simply horsecars retrofitted with motors, electrical controls, and trolley poles. These were generally short-lived, not least because a roof designed solely to keep passengers under cover was too flimsy to support the weight of trolley poles. But lightweight cars were wanting durability in many other ways, and it was clear that larger passenger loads demanded sturdier cars. The first step in this direction was the development of a separate metal "truck" that supported the weight of the carbody through coil springs, often with elliptic (leaf) springs extending from the ends of the truck.

But there were inherent drawbacks to single-truck cars. First, the length of the wheelbase (that is, the distance between the axles) was limited by the 90-degree turns that trolley lines often made from one downtown street to a cross-street; if the axles were spaced too far apart, the wheels could not follow the tracks. Second, because the length of the truck was limited, but not necessarily the length of the car itself, a 4-wheeled trolley with large overhang would "gallop" from side to side if the motorman tried to go much faster than 15 miles an hour. Some cable cars were almost that fast.

As galloping limited speeds, as the need for greater capacity became increasingly obvious, and as new and longer trolley lines were opened at the turn of the twentieth century, the design of streetcars veered back toward the design of railway cars, as in the beginning. More and more of them were built with a 4-wheeled truck at *each end*, and with the weight carried on the two "kingpins" on which the trucks swiveled. (Overall, nothing was more important than the kingpin, and this expression quickly found its way into common discourse.) An 8-wheeled "double truck" trolley had two advantages; it could transport more passengers and it could transport them faster. Bobtails all but disappeared. Rather than having open platforms, double-truckers were designed with enclosed vestibules. But the routine at the end of the line remained unchanged: the conductor would raise one trolley pole and lower the other, flop the hinged seat-backs over so that riders would be facing forward if they got a chance to sit down, and then trade stations with the motorman.

In Great Britain and in most of Europe, a boarding passenger normally expected to be able to find a seat at any time, even in rush hour. This was never the case in the United States. By the turn of the century, a typical American double-truck streetcar had a capacity of about a hundred riders, but had seats for only forty. The aisles were for standees, who could grab hold of overhead straps to help them keep their balance (hence the expression "straphanger"). A trolley would stop and start again and again, and passengers obliged to stay on their feet for a long way were rarely happy passengers. The people who owned trolleys, on the other hand, had a saying among themselves: because the fare was almost never more than five cents, "the dividends are in the straps." Although a gentleman was expected to yield his seat, and usually would, ladies increasingly came to regard riding in trolleys as unpleasant, or even worse. The operating companies were continually accused of stretching out headways, the scheduled time between one car and the next. In the "off-peak" period when most riders were on shopping trips or running errands there might be a long wait, and during rush hours every car would inevitably be filled to overflowing.

Whatever the disagreements about crowded cars and rider comfort, large double-truck cars were not entirely satisfactory to the men who owned them either. Some dilemmas seem to come and go and then come back, and with streetcars one such dilemma was weight. Too heavy? Too light? Too heavy? Here, that pendulum would swing once again. In the 1840s and 1850s, the earliest horsecars weighed almost as much as railroad cars and they overtaxed the horses, which were valuable property. In the 1890s, some of the earliest trolleys were so lightly constructed that they would shake themselves to pieces after a short time in daily service. So the trend at the turn of the century was toward double-truck cars, larger and larger. But the people who owned and operated streetcar lines quickly discovered that large double-truck trolleys consumed much more power—a significant tradeoff—and, as importantly, they were hard on the track. They would batter and eventually split the railhead, especially at stops. Rails could be replaced, of course, but not without a lot of work digging up the street and then repaving it. So weight was already a concern even as the materials from which streetcars were built began to change and they would tend to get even heavier.

STREETCAR MATERIALS

Until well into the twentieth century, streetcars were assembled almost entirely of wooden components—even the basic structural members running the full length of the car, the sills. The only exceptions were fastenings—nuts and bolts, screws—and of course the trucks, motors, and running gear, made of both iron and steel, and with some copper and brass as well. But as the steam railroads ordered ever-increasing quantities of steel rail—as the lands were "welded together," in Walt Whitman's phrase—they also spurred the steel industry to new levels of productive capacity, which meant that prices were falling even as so-called prime trees were fast disappearing from America's forests and hardwood was becoming more costly. The first steel railway cars that were not considered experimental were delivered in 1904 and 1905 by the American Car & Foundry Corporation to the Long Island Railroad and to August Belmont's Interborough Rapid Transit, New York's first subway. (Since a 1903 catastrophe in the Paris underground in which dozens of people died, there had been serious concerns about fires in subways.) During the next few years, as railways were ordering more and more steel coaches and sleeping cars, orders for rolling stock from urban transit firms also began specifying steel components in the interest of economy.

First it was the sills and then side panels and the supports for the roof. By riveting side panels and structural framing together, a car builder would create a girder that could support the weight of the other components. Even though they still had wood in the roof, flooring, and window frames, most streetcars built after 1910 were primarily steel. While the venerable Stephenson factory built wooden cars in New Jersey to the very end in 1917, there was little doubt about steel cars being less expensive. Yet double-truck cars had grown very heavy even when they were made of wood, and steel cars weighed even more. People in the street railway industry—it was by now an industry for sure, worth billions of dollars—would keep thinking about lightweight designs as profit margins began to narrow ever so slightly as war broke out in Europe and it seemed more important to keep an eye on power consumption and the cost of renewing trackwork. At annual meetings of the American Electric Railway Association, there would even be talk of reviving single-truck cars and making up for their lack of capacity by running more of them—accepting that tradeoff—as we shall see.

STREETCAR MANUFACTURERS

Narrowing profit margins were of no real concern to the men who controlled trolley systems during the first years of the twentieth century, certainly not the enormous networks in big cities. They felt sure that any downturn would be only for the short term, and they still looked to increases in patronage and profits that would continue indefinitely. A hundred years later, when mass transit systems can typically recover only a fraction of their operating costs at the farebox, it is hard to recapture a sense of how much wealth and power traction magnates really had. It is hard to imagine urban streetscapes dominated by endless processions of streetcars, hard to imagine the vitality of an industry whose dollar value made it the fifth largest in the United States. And, if it is hard to imagine how big the business of operating streetcars was, it is likewise hard to imagine the commanding presence of the factories that manufactured rolling stock. Some of the operating companies themselves got into manufacturing on a large scale. The Los Angeles Railway was a notable instance of this, turning out hundreds of streetcars at its South Park Shops on Central Avenue. So was the company in the Twin Cities at its Snelling Avenue Shops in St. Paul, which manufactured not only streetcars used in Minneapolis and St. Paul but also in Seattle and Tacoma, Chattanooga and Nashville. These cars were amazingly durable, even if one must take with a grain of salt a company claim in the 1950s that Twin Cities patrons "never realized that they were riding on trolleys that were 40 to 50 years old" (Kieffer 1958, 25).

Though a number of other street railway companies built their own cars, most of them solicited bids from industrial firms that specialized in rail-borne conveyances. For many years the giant of the streetcar manufacturing industry was the firm founded by John Stephenson, but after the advent of electricity Stephenson was eclipsed by three other companies: the St. Louis Car Company, the Pullman Car Manufacturing Company in Chicago (both of which also built rolling stock for steam railroads), and the J. G. Brill Company of Philadelphia, which specialized in vehicles for electric transit, although it made railroad cars, and even trucks and buses, as well. John George Brill (1817–1888) had spent 25 years working for a Philadelphia manufacturer of railroad cars before founding his own firm in 1868 along with his son G. Martin Brill (1852–1906). The Brills produced horsecars and cable cars in the 1870s and 1880s, but their business took off with the advent of trolleys—Frank Sprague used Brill cars in Richmond—and the enormous growth in patronage and profits in the 1890s.

At the turn of the century, the Brills set out to dominate streetcar man-ufacture, absorbing the Stephenson firm and others in St. Louis, Cleveland, and Springfield, Massachusetts. This was just at the time when other great industries from steel to shipping were being consolidated into "trusts," and a streetcar manufacturing trust may have been what the Brills had in mind. That degree of dominance eluded them, but their firm did become much the largest in the world, total production over the years amounting to more than 45,000 vehicles. In 1912 Cie. J. G. Brill was established in Paris. Even-tually Brill produced streetcars and streetcar components that were sold in all parts of the United States, in every Central American and South American country, in most of the countries of Europe, in Burma, China, India, Japan, Thailand, Zanzibar—and in every other country, or so it seemed to many contemporaries (Charlton 1957, 13). Brill's "package type" selling sounds to modern ears a lot like what IBM's was at one point. Brill held patents on dozens of different components, from trucks to trolley wheels, seats to gongs. To buy a Brill car was to get the works—everything, so the expression went, but the men to run it.

TRACTION MAGNATES

A business enterprise that achieved the vitality of the J. G. Brill Company is suggestive of the overwhelming importance of the street railway industry in many countries, but particularly in the United States. Before the advent of automobility, street railways flourished in every city and every town of any size at all. Streetcar motormen and conductors were among the major occupational groups tallied by the U.S. Census. In big cities, streetcar systems were enormously profitable and the capitalists who controlled these systems not only became enormously rich, they often exerted unprecedented control over municipal politics. City officials granted them all sorts of special privileges, such as franchises that would allow them to monopolize mass transit for 99 years, or sometimes even 999 years. The Widener Library at Harvard University, the Yerkes Observatory in Chicago, and the Huntington Library and Art Gallery in California all stem from riches accumulated by traction magnates. In their own time, however, men like Peter Widener, Charles Tyson Yerkes, and Henry Edwards Huntington were often viewed with disfavor, as one of the most conspicuous sorts of "robber barons." They were thought—often with good reason—to be set on manipulating public trust, buying influence, exploiting labor, and neglecting the best interests of their patrons, even abusing them.

In a period of less than a decade, Widener and his associates William Whitney and Thomas Fortune Ryan made $100 million from New York's Metropolitan Street Railway. Yerkes (1837–1905), the robber baron *par excellence*, had actually served time in jail for embezzlement in Philadelphia before coming to Chicago in 1881 and forming a syndicate that developed a tactic for reaping millions in street-railway profits "without investing a dime of its own money." Yet his syndicate was also responsible for completing the "Loop," which brought all the elevated lines together downtown, along with electric railways running north, west, and south, and was "probably the single most important mass-transit project ever built in Chicago"—and is of course still a Chicago landmark (Young 1998, 51, 61). After a furor over his illicit attempt to multiply the value of his franchises, Yerkes left Chicago and engaged J. P. Morgan in a struggle for control of rights to the London Underground. A largely accurate depiction of Yerkes was Frank Algernon Cowperwood, the dominating character in Theodore Dreiser's trilogy, *The Titan*, *The Financier*, and *The Stoic*. A former Chicago mayor who had been Yerkes's bitter enemy later called him "gallant though perverted."

The ambitious and sometimes unscrupulous men who consolidated streetcar lines occasionally sought to penetrate old neighborhoods where there had been none and where they were resisted with the sort of arguments that today we associate with NIMBY ("not in my back yard") sentiments. For example, upper-crust citizens in Montclair, New Jersey, exploited fears that their town would become "a dumping ground for Dutch picnics" (Middleton 1987, 79). Fifth Avenue never did have trolleys, and the bus fare was 10 cents, twice the fare anywhere else in New York City. But efforts to limit patronage to the upper crust and to keep tracks off city streets were limited to only a few places besides Fifth Avenue—along the Chicago lakefront for one. Almost everywhere else the expansion of street railways was regarded as a municipal boon even as the men who controlled them grew richer and lost favor in the eyes of the public. Between 1890 and 1902, mileage in the United States increased from 5,783 to 16,652 (all but 422 miles of which were electrified), and the number of streetcars almost doubled, from 32,505 to 60,290.

RECREATION AND ENTERTAINMENT

Civic boosters saw street railways as a prime engine of economic development. First and foremost, they enabled citizens who could afford suburban living to commute (literally "to travel regularly between one's home and one's place of employment") over a distance of several miles. But they also served many functions that were not expressly utilitarian. There were lines running out beyond the suburbs to the countryside, to cemeteries (there were special funeral cars), or to scenic sites for picnics and outings. A 1907 census counted 467 "parks and pleasure resorts" owned or operated by street railway companies. Trolleys often provided transport to an entirely new kind of "pleasure resort" called an amusement park, with merry-go-rounds, roller coasters, and various other sorts of "rides." Brooklyn's Coney Island became the most famous, but other notable examples included Kennywood Park, operated by the Monongahela Street Railway Company, 12 miles from Pittsburgh; Playland by the Sea, at the end of the Geary Street line in San Francisco; and Glen Echo Park, at the end of the Washington Railway & Electric Company's picturesque line running from Georgetown northerly along the palisades of the Potomac River.

Of course the trolley was a "ride" in itself. Cars without any window-panes and sometimes without sides at all offered cool breezes in summertime (Figure 2.4). Courtships in which trolleys played a central role became the stuff of storybooks. In the movie musical *Meet Me in St. Louis*, Judy Garland sang about the sound of the two different bells rung by the motorman and the conductor:

> Clang, Clang, clang went the trolley,
> Ding, ding, ding went the bell. . . .

In cartoonist Fontaine Fox's "Toonerville Trolley," the "Skipper" shuttled riders back and forth to the train station in Pelham Manor, New York.

Figure 2.4: Open cars were boarded from the longitudinal step, from which the conductor also collected fares. This car, obtained second-hand from a Connecticut company by the Cooperativa de Transportes Urbanos y Sub Urbanos de Vera Cruz, operated year round. Cars similar to this one, but with a center aisle, were called "convertibles" and had side panels and windows that could be removed or replaced, depending on the season. (Author's photo)

The trolleys of Los Angeles and Hollywood became familiar worldwide because they appeared in the movies so often—especially in the comedies of Harold Lloyd, Buster Keaton, and the Keystone Kops. Booth Tarkington's *The Magnificent Ambersons*, which chronicles a family whose "splendor lasted throughout all the years that saw their Midland town spread and darken into a city," begins with "the little bunty street-cars on the long, single track that went its troubled way along the cobblestones."

THE INTERURBANS

In 1914, when what became known as the Great War broke out, there were trams everywhere in Europe and throughout the British Isles. In the United States, trolleys were totally interwoven into the fabric of urban life—and they were a big part of rural life, too, because in many parts of the country interurban trolley lines connected cities and towns. With a smaller investment in construction and lower labor, power, and maintenance

Figure 2.5: The Chicago, North Shore & Milwaukee was unusual among interurban electric railways in operating not only single cars but unitized trains, called Electroliners, articulated so they could negotiate the tight turns of the Loop in Chicago. In Milwaukee, they ran right through residential neighborhoods. The North Shore was one of the last interurbans to stay in business; after it was abandoned in 1963, the Electroliners got a second life with a line running between Philadelphia and Norristown. (Author's photo)

expenses, an electric railway was able to schedule more frequent service with single cars and 2-man crews than a steam railroad could do with locomotives, trains, and a standard crew of five men; cars could stop at country crossroads and they could run down Main Street to a terminal right in the center of town, often sharing the same tracks used by the local trolleys (Figure 2.5).

Just as streetcars were integral to the fabric of urban life, they had a likewise profound impact on the countryside:

> For the first time, a farmer's wife was able to go into town, shop for an hour or two, and return to the farm by dinner time. A resident of a country town could spend part of a day in the city and return home at his leisure. The salesman who had been able to visit one or two towns in a day could now visit four or five. Even on the busiest railroad lines in the Midwest locals rarely ran more often than twice a day, but an interurban might schedule 16 cars a day (Hilton 1960, 91).

Interurban lines converged from every direction on the great Indiana Traction Terminal in Indianapolis: from Dayton and New Castle to the east, Muncie and Ft. Wayne to the north, Terre Haute to the west, Louisville to the south. In Ohio, no town larger than 10,000 lacked electric interurban connections, and total mileage statewide was nearly 2,800. Los Angeles was the hub of the Pacific Electric, a 1,100-mile interurban system whose "Big Red Cars" skirted mile after mile of sandy shoreline, swept past endless acres of orange groves, and climbed into the foothills of the San Gabriel Mountains. The PE even reached a resort called Mount Lowe that was established by Thaddeus S. C. Lowe—famed for his balloon reconnaissance on behalf of the Union Army during the Civil war—via a funicular, a system comprising two cars counterbalancing one another on a steep incline (Figure 2.6).

CLOUDS ON THE HORIZON

Under control of the Southern Pacific Railroad after 1911, the Red Cars remained a completely separate operation from the local streetcar system in Los Angeles—the two did not interchange transfers and even the track gauge (the distance between the rails) was different. So, too, the lines that converged at the Indiana Traction Terminal were distinct from the local streetcar system in Indianapolis. Even without counting the interurbans, by Armistice Day in 1918 local trolley lines in the United States represented an investment of $4 billion and yielded their stockholders profits of $600 million annually. The effects of wartime inflation were starting to become obvious, and some of the interurbans were turning out to be a poor investment after only a few years, as rural roads were improved and motor vehicles became increasingly popular among farmers. And yet the men who had invested in the unified big-city systems—with their thousands of streetcars

Figure 2.6: Funiculars like the Mt. Lowe line were built in several cities, including Cincinnati, Duluth, and Pittsburgh, and on the Palisades across the Hudson River from Manhattan. Two Pittsburgh funiculars remain in operation to this day. In Los Angeles, Angels Flight operated between 1901 and 1969, climbing a 33 percent grade for 335 feet to an increasingly seedy neighborhood called Bunker Hill. Its noir ambience was woven into the popular imagination through the movies. Jimmy Stewart, Lon Chaney, Ann Sheridan, and William Holden all had their dramatic moments around Angels Flight. (Author's photo)

and their lines crisscrossing every downtown street—showed no concern. As they saw their own situation, franchises entailed some obligations, but much more importantly they guaranteed future earnings, and the speed and range of trolleys promised perpetual gains in patronage and revenue as urban populations grew into new suburbs. What they anticipated are what we now call windfall profits.

Somewhat like the masters of Enron and WorldCom at the turn of the twenty-first century, some traction magnates had shamelessly "watered" their stock—that is, boosted its par value without doing anything to enhance the real value of their property. But many of them did not need to do anything like that. What they had, in an expression of that day, was a cash cow. Patronage had skyrocketed during the first decade of the twentieth century and increased continuously during the second; except during the war, monetary deflation had prevailed for even longer than that, and traction magnates had never seen any indication that debt could not be covered with farebox receipts that would always grow. Nationwide, streetcar ridership had increased from 2 billion in 1890 to 5 billion in 1902 (160 percent) to 11 billion in 1912 (120 percent). Then it grew to 13 billion in 1917 and 14 billion in 1923. This was slower growth than before, but what nobody knew at the time was that it would not reach that level again for 20 years, and even then it would only be a temporary wartime spike.

3

Competition

The debut of the trolley in the 1880s signaled the end of the line for horsecars and cable cars. Something else happened at almost exactly the same time that the trolley appeared; so did the automobile with power by internal combustion, that is, with the fuel (gasoline) being ignited inside the engine itself rather than in a separate firebox, as with a steam locomotive. Automobiles—or in a more general term that takes in commercial uses, motor vehicles—would eventually signal the end of the line for trolleys, just as trolleys had done for horsecars and cable cars. But at first automobiles were not perceived as a means of everyday transportation at all. Rather, "they were all sports cars in the sense that speed, adventure, and the appeal of technological novelty were among their key attractions" (Volti 2004, 9). Even when they were reconceptualized as something more than "rich man's toys," their utility was limited by the poor quality of "thoroughfares designed to move traffic through ambient surroundings." In cities and towns there were paved streets, of course, but they served a different purpose: allowing local access to shops, schools, churches, hotels, and civic buildings. They were not designed to accommodate motorists who simply wanted to get back and forth from their suburban homes (Mom 2005, 745–72).

As more tax dollars were made available for thoroughfares designed to "move traffic through ambient surroundings," more people began to drive "autos" (literally, to provide their own mobility) rather than use transit (a word that implies a passive relationship to a conveyance). Sales in the United States jumped from 24,000 in 1905 to 896,000 in 1915, and then the numbers went out of sight in the 1920s as prices fell within general reach. But long before automobiles were widely affordable, everybody was just dying to take a ride. Even in 1905, when not everyone had even seen an auto as yet, the year's most popular song had been "Come away with me Lucille . . . in our merry Oldsmobile."

JITNEYS

Prime evidence of this enthusiasm was the so-called jitney craze that began in Los Angeles in the summer of 1914, just as world war broke out in Europe. A jitney could be anything from a Model T phaeton to a "motorbus" (a new coinage, like "omnibus" nearly a century before) that could accommodate a dozen or more passengers. The common denominator was an enterprising driver who followed the same routes used by streetcars and picked up passengers, sometimes delivering them exactly where they wanted to go for the same price as a trolley ride (jitney was slang for a nickel, five cents). By 1915, jitneys had spread to the entire nation. Estimates of the total number varied wildly, though the newspapers at one point reported that there were more than 60,000 of them. Actually it was almost impossible to make even a rough count, but it was obvious that there were swarms of jitneys in many different parts of the United States. In just 1 year, so said the general manager of a transit firm whose trolley lines blanketed northern New Jersey, jitneys had "skimmed off $4 million worth of business."

While it was easy for someone in the trolley business to overstate the difficulties he faced, real tallies could be made of layoffs caused by the downturn in revenue; the Los Angeles Railway, for example, released more than a hundred men who had been employed to refurbish some of its oldest cars at its South Park Shops. In many places there were rumors about plans to cut the workforce by redesigning cars so that motormen could assume the duties of conductors, and the topic of "one man operation" could be counted on to inflame controversy every time it came up.

In hundreds of cities and towns, streetcar companies sought to have jitneys outlawed as unfair competition. Unfair, because trolleys, as one official put it, "were subject to rigid regulation . . . they pay heavy sums into city and state treasuries in the way of taxes and license fees, and as responsible corporate bodies they can be held to account for their claims" (quoted in Bail 1984, 10). Jitneys came and went as the men who drove them pleased; they were subject to no regulation, paid no fees, and were held to no account. It was well understood that municipalities benefited from the paving assessments that trolley companies were obliged to take care of, and there was good reason to enact ordinances restricting jitneys or banning them outright. In a special election in June 1917, voters approved an ordinance to exclude jitneys from a 40-square-block area in downtown Los Angeles. The tally was 52,000 to 42,000, but the outcome was clearly affected when the Los Angeles Railway and the Pacific Electric both "declared a general holiday and made 550 autos available to get voters to the polls" (Swett 1951, 44). Sometimes, municipal officials were tempted to *encourage* wildcat competition as a way of getting back at "corporate bodies" that had not always seemed terribly responsible; the southern California beach town of Santa Monica gave a green light to the operation of jitneys that competed directly with the Pacific Electric. There was, after all, a lot of pent-up resentment about the excesses of traction magnates, and a feeling that the "trolley trust" still needed taming, not commiseration. Besides, most people thought that jitneys were fun:

Hush, little Ford,
Don't you cry,
You will be a jitney
Bye and bye.

INFLATION

Whether or not jitney competition was restricted or banned, the burgeoning popularity of motorcars foretold hard times for mass transit. So did a changing economy. For nearly 50 years after the Civil War prices were stable or even declining; raw materials cost the same amount, wage rates remained the same year after year. Since the beginning of the world war, however, that had changed drastically, and everything was becoming more expensive. Steel rails cost 50 percent more, but most important was the cost of labor, which was by far the largest day-to-day expense for street railways. Motormen and conductors were hired from a large surplus pool of itinerant labor. Although pay rates were about the same as for semi-skilled work in manufacturing, pay on an annual basis was actually lower because motormen and conductors worked longer hours, often 14 hours a day, 6 days a week. Turnover was high, morale perpetually low. And there was a collateral problem: when patrons were confronted with operating employees who were not happy with their lot, they would focus their resentment on management, not the working men.

As inflation drove consumer prices to nearly double the level they had been before the war, labor unrest grew, and President Woodrow Wilson was compelled to establish a labor board to arbitrate pay disputes and to set wages that were sufficient "to insure the subsistence of the worker and his family in health and reasonable comfort." With street railway workers, the board ultimately adjusted wages not only to compensate for inflation but also to gain parity with wages in manufacturing. By 1920, conductors and motormen were often earning twice as much as before. Moreover, large segments of the labor force were now unionized—in many cities, all the operating employees—and this insured that there would be continued upward pressure on wage rates even as street railway profits were leveling off and most companies were finding their capacity to secure loans for expansion and improvement severely constricted.

Cities were continuing to spread out into new suburban neighborhoods that were beyond "the end the line" for streetcars, and independent bus operators (the name preferred by people with jitneys) were anxious to fill "gaps" in service. "People living off the car lines want service brought closer to their doors," said an executive with a Detroit motor vehicle company, who went on to warn that streetcar companies had "better inaugurate bus service of their own" if they were not going to extend their tracks. Faced with unregulated jitney competition in many of the towns it served, the Pacific Electric was a pioneer in this regard, with a bus line in San Bernardino as early as 1917 and another in Redlands in 1921. But when transit firms floated trial balloons about raising fares in order to accumulate capital for extensions, whether trolley or bus, they met with unremitting hostility from the public, from municipal officials, and often from the press. It was hard to believe that an enterprise that had yielded its owners such riches in the recent past could have been seriously affected by inflation—that it could actually be short of cash. Along with proposals to cut operating expenses by instituting one-man operation, the subject of fare increases became a constant bone of contention.

FARES

The dispute over fares in Los Angeles can serve as an example of a conflict that arose time and again elsewhere. The Pacific Electric was headquartered there, with major terminals on Main Street and Hill Street and 4-track main lines on both its northern district (connecting with the San Gabriel Valley and points east) and its southern district (connecting with the harbor and points south). Operations spread over a vast hinterland, as far to the east as Riverside and Redlands, as far to the south as Huntington Beach, Newport, and Balboa. The PE also operated local streetcar systems in many of the outlying cities and towns it served such as Long Beach, Pomona, and Pasadena. But local service in Los Angeles itself was the business of another company, the Los Angeles Railway, which was owned by the same multimillionaire who had once owned the PE, Henry Huntington. In the world of letters, Huntington was known as the world's foremost collector of rare books, but in the popular imagination he was the man who had bought and sold "all the unimproved real estate in Southern California" (Thorpe 1994, 202).

The Los Angeles Railway was a huge enterprise, as intensively focused as the Pacific Electric was expansive, and it actually carried far more passengers. There were 370 miles of track, 16 electrical substations, a 16-acre shop complex, 1,200 streetcars and a growing fleet of buses, carbarns that stretched as far as the eye could see, and a 10-story office building. There were lines on nearly every downtown Los Angeles street and trolleys were "through routed" from one outlying terminal to another in order to avoid looping downtown and risking gridlock. Depending on the relative consideration given cost of reproduction as against original cost, estimates of its worth ranged from $30 million to $45 million and even higher. In 1918, after 4 years of wartime inflation, the company's rate of return had dwindled to 2.8 percent. The situation improved in the early 1920s as costs leveled off, but in 1925 profits again began to ebb in the face of more inflation and declining patronage, the result of a seemingly unquenchable enthusiasm for automobiles that had been signaled by the jitney craze. Ten years earlier, few Angelenos could afford to own an automobile, but now autos were becoming affordable, especially with a burgeoning second-hand market. Even as the population kept growing at more than 10 percent a year, faster than any other American city (it would reach 1,238,048 in 1930, a 118 percent increase during the decade), streetcar patronage was not keeping pace: more and more Angelenos, fewer of them riding trolleys.

There had always been fewer riders at midday, but now the trolleys might be nearly empty between 10:00 a.m. and 3:00 p.m., in between the so-called rush hours. More people were *driving* downtown instead of taking the trolley, especially middle-class women on shopping trips or families aiming to dine out or go to a movie. Women had rarely been fans of streetcars. Waiting in a safety zone with traffic whizzing by inches away was not pleasant. Neither was climbing the steps of a streetcar in a skirt. Neither were the men that one was likely to encounter once inside, with their cigar smoke, chewing tobacco, and coarse language. If women had an option, and more and more of them did, they were strongly motivated to take the wheel—to leave home when they wanted and to drive exactly where they wanted, in privacy, and, on a deeper level, "in search of personal and political freedom" (Scharff 1991, 89).

The convenience of driving was hardly less appealing to middle-class gentlemen commuting to work from the suburbs. Even as early as 1923, nearly half the people entering and leaving the central business district during a typical 12-hour weekday did so in autos; by 1931 it was 62 percent.

Why had trolleys attracted so many patrons just 10 or 15 years before? It was because so few people had an option. Nothing could beat a nickel carfare for a ride that could be several miles, certainly not what it cost to drive an automobile and park in a pay-lot downtown. But when an auto could actually be fitted into the family budget, people started taking more into consideration than just the *cost* of travel. There were less tangible matters; there was the expression that captured popular fancy, "riding in style." Time and again Angelenos heard that trolleys were "hopelessly out of date." One of the newspapers insinuated that "influential people" would want to drive wherever they went, while the working class would be better served by new buses, perhaps municipally owned and not under the control of someone who had never seen the inside of a regular streetcar. The same arguments about outdated trolleys and nice new buses could be heard in many other cities, most notably in New York. But there the 5-cent fare was considered to be sacred (with the one exception of Fifth Avenue) while in Los Angeles this was not so certain. The company had sought permission to increase fares once before and then backed off when profits started to recover, but this had caused a great deal of concern among patrons. Despite all the growing popularity of autos, hundreds of thousands of people still found trolleys indispensable.

A FAIR RETURN ON INVESTMENT

So, imagine the response in November 1926, when an attorney for the Los Angeles Railway went before the California Railroad Commission and asked for permission to increase the fare to 7 cents or four tokens for a quarter (street railways always sold tokens at a discount). The nickel fare no longer yielded a fair return on investment? Ha! The company seemed to have been doing just fine in 1922, when the return was more than 9 percent. So what if it was only 4.6 percent in 1926? Why should not returns be averaged over a period of several years? And so what if streetcar firms in other cities had been permitted to raise fares? Operating costs in L.A. were as low as any in the nation, franchises entailed no annual fees, power was cheap, maintenance less complicated because of the mild climate, and the payroll easier to meet (Los Angeles being staunchly "open shop" [nonunion], the Los Angeles Railway's conductors and motormen were paid about the same wages as prevailed in the Deep South).

Accepting the contention that returns were reasonable if averaged over a period of several years, the Railroad Commission denied the company's plea. There would be no voyage on to what one imaginative commissioner termed "the uncharted sea of multi-coin fares" (quoted in Post 1970, 286). But there was an appeal, the case eventually went all the way to the U.S. Supreme Court, and there the company won. Seven cents would be the fare (Figure 3.1).

At just about the same time, in early 1930, the court ruled in a Baltimore case that in order to escape the charge of confiscating private property (that is, not permitting fair return on investment), a state regulatory agency had to "allow" a streetcar company to earn a 7.5 to 8 percent profit on its assessed valuation. At that point, however, such profits were far beyond reach, in Baltimore or in Los Angeles or anywhere else. Unemployment was at an all-time high and trolley patronage was falling month after month. The depression had drastically affected the "off-peak" patronage—shoppers in the mid-morning, theatergoers in the evening—that was essential to keeping fleets of streetcars, and the men who ran streetcars, from sitting idle between rush hours. Yet, even if it was reasonable to assume that the depression would not last indefinitely, one of mass transit's afflictions was to prove both chronic and degenerative, the growing popularity of automobiles.

The situation in L.A. was not unique. Even with the economy in the worst slump ever, more autos were crowding downtown streets nearly everywhere in the United States. In big cities this meant increasing congestion, and *The Saturday Evening Post*, a popular weekly magazine, had coined a new expression, "traffic jam." Nobody liked traffic jams, but they affected trolleys as much as automobiles (worse, actually, since trolleys were not maneuverable), and more and more private citizens seemed to prefer to wait out delays in their own car, rather than stuck on a trolley surrounded by people who might not be pleasant to be around.

As for the people responsible for managing transit firms, they were trapped in a no-win situation. They could compete with motorcars only by trying to match their speed, comfort, and convenience. This might not actually be feasible under any circumstances, but any serious attempt to do so would require extending lines into newly built-up neighborhoods, acquiring modern equipment, and generally upgrading service and catching up with deferred maintenance.

Another Financial Debacle!

Figure 3.1: The Los Angeles Railway had been charging 7 cents while its appeal awaited the Supreme Court decision, and passengers were given vouchers worth 2 cents. If the company lost, the money would have to be refunded. Because the ruling took 14 months, many Angelenos had 20 or 30 dollars riding on the outcome. Mass transit was an old standby on editorial pages everywhere, and the staff cartoonist for the *Los Angeles Times* depicted the disappointment.

Higher fares might help pay for improvements, at least as an abstract proposition. But whenever fares did go up, the gain in net revenue was hardly worth the "unwanted result," a residue of public ill will. As the depression persisted, transit management was growing desperate to increase income somehow. Trying to raise fares was problematic. So was trying to cut payrolls through one-man operation, though that tug-of-war between labor and management would persist for decades. But there was a third option, simple if not necessarily inexpensive: a way of deriving more income from the same number of riders paying the same fare, by improving the efficiency with which fares were collected.

SINGLE-ENDED CARS

Except for bobtails, which proved to be a short-lived phenomenon for the most part, almost all trolleys were designed to have a motorman at the controls and a conductor to collect fares. Almost all were double ended—that is, they had the same kind of platforms at either end and could run in either direction. At the end of the line, the conductor would swap positions with the motorman, who carried his controller handle from one end of the car to the other. Motormen insisted that they had their hands full just operating their car safely in heavy traffic, especially when the tracks made a right turn directly in front the automobiles in the right-hand lane. Distractions inside the car, making change, collecting fares, transfers, keeping order: all that was up to the conductor. Because it was essential to minimize "dwell time"—the interval when a streetcar had to remain stopped in order to get everyone aboard—passengers would board both front and rear. The conductor would range through the car, calling out "Fares, please!" and ringing up each new fare on a device called a fare register which usually had a series of levers running the length of the car. Conductors on busy lines would handle vast quantities of coinage. Management assumed the worst, that conductors would "knock down" by pocketing some of it, and employed spotters to report whether they were doing the right thing. Still, in a streetcar with more than a hundred people crowding aboard, even the most conscientious conductor, even if he had a photographic memory, could easily fail to single out every new passenger mingling (sometimes with deception in mind) with those who had already paid.

Hence management's growing incentive to implement some more efficient system of fare collection even though this could necessitate single-ended cars, cars designed differently in front than in back. Rather that walking the aisle, the conductor would stay put and customers would be obliged to hand over the fare as they passed him by. (And it was "him," always, right up until labor shortages during World War II resulted in women sometimes being hired as conductors and even as motormen.) On busy lines the big problem was dwell time. Somehow, passengers would still need to be able to board quickly so a car could stop and go without delay; there could be no waiting lines outside in the safety zone. Beginning in the 1920s, a number of new designs were put into service that, in theory at least, would ensure that every rider paid while not extending dwell time. One way or another, part of the car was turned into a "prepayment" area. With the "Pay as You Pass" car, it was the entire front half. Passengers would board in front, and the conductor was stationed just inside a pair of doors in the center, to collect

Figure 3.2: Front-entrance, center-exit "Pay as You Pass" cars like this one were named for Peter Witt, a Cleveland street railway commissioner who developed the design. Besides Cleveland, there were fleets of Peter Witts in Brooklyn and Baltimore—but none any further south, where the configuration of streetcars was inevitably influenced by measures to keep them racially segregated. This second-hand Peter Witt car is seen operating in Mexico in 1961. (Author's photo)

a fare either as they exited or moved to the rear of the car to continue their ride and exit later (Figure 3.2). Then, there were various types of PAYE cars ("Pay as You Enter"); in one type, the conductor collected fares on an oversized rear platform while the front door was reserved for getting off. Another was called a "Near Side" car and had two front doors, one for entering and one for exiting, and the conductor was stationed behind the motorman.

PLATFORM COSTS

Every operating official believed that any front-entrance car offered a potential revenue gain much greater than merely assuring that everyone paid their fare. There *could* be a 50-percent saving in labor costs because such a car could be operated without any conductor at all: one man, the motorman, could attend to fares as well. Or so management insisted. Those who worked on the cars saw the matter quite differently. Whenever they were unionized—in about half the cities and towns where streetcars operated— they would instantly fight back any proposal for one-man operation, even when it was depicted by management as a matter of dire financial necessity. Time and again, conflict over one-man operation roiled streetcar labor and led to work stoppages, strikes. And the effect of a strike was always to reduce patronage in the long run; after it was settled, there were never as many customers as before. The smaller the city, the shakier the financial situation, and the more weakly organized the workforce; hence the more likely that signs would appear next to the front doors of streetcars that said *Have Fare Ready* or something of the sort which meant that riders should expect to be paying the motorman, not a conductor. It usually took a municipal ordinance to authorize one-man operation, and no issue was more contentious, but this became a cause to which transit management was steadfastly dedicated. A 50 percent reduction in the cost of operating manpower might not affect the bottom line like a 50 percent increase in patronage, but it would surely help.

The term that transit firms used for the manpower required to operate streetcars was "platform costs," and everyone understood it: There often were signs inside streetcars saying *Do Not Talk to the Platform Men.* The readiest means of reducing platform costs was simply to increase headways—that is, to schedule service less frequently and have fewer cars out on the line at any given time. This was indeed a way of reducing manpower that was less obvious than one-man operation, and yet there was nothing more certain to get customers steamed up than a long wait for a streetcar that was already crowded when it finally came along. "Service?" Could it be called service when people were "compelled to stand for miles at a stretch," as one complainant put it, "often herded together like cattle or sheep in a stock car" (quoted in Bottles 1987, 73)?

So, what if the cars were larger? Double-deck cars were used almost everywhere in Great Britain (Figure 3.3). But double-deckers rarely caught on with transit firms in the United States, and seldom in France or Germany, for that matter. Too slow to load and unload, that was the usual explanation—and it was certainly true that on busy lines they would affect dwell times negatively—but "national styles" in technology usually have complex explanations and this has often been the case with urban transit. Occasionally there was interest in another high-capacity design consisting of two articulated car bodies with one truck between them, and in a few cities two trolleys would be coupled together, or a trolley coupled to a trailer with no motors. For riders, these might provide a better chance of getting a seat, but they skirted the "platform costs" problem as management saw it.

The two-man crew was a tradition that went back to horsecar days, when there was a "driver," before there was a motorman. Then, just as trolleys emerged on to the main streets of America in the 1890s, the population of cities and towns began to grow and disperse as never before,

Figure 3.3: A London Transport double-decker, of which there had been thousands in the 1920s, is seen here just a few months before its replacement by trolley buses in 1938; the twin overhead wires are already in place. (Henry Priestly photo, courtesy National Tramway Museum, Crich, Derbyshire)

and there were huge increases in patronage. Having a conductor to collect fares seemed more necessary than ever before, and it seemed essential when high-capacity double-truck cars began to appear in large cities after the turn of the century. Single-truckers were still used in many places, however, and where labor was unorganized they looked like ready candidates for one-man operation. All it took was sealing the rear doors (or designating them for exit only) and requiring that the motorman mind the riders along with his other duties. But management's push toward one-man operation would remain divisive right up to the time that trolleys disappeared in favor of buses, which almost never had more than a driver. Indeed, choices were often driven as much by the one-man issue as by rational assessments of competing technologies.

BIRNEYS

In the years just before the war, Henry Ford was showing the world how the cost of manufacturing a vehicle could be vastly reduced by using interchangeable parts. When the Model T was introduced in 1908 it was handcrafted and priced at $850; in 1913, after Ford shifted to assembly-line production in Highland Park, Michigan, the price fell to $350. This was largely a matter of deploying manpower in a series of highly specialized tasks: The worker who put in the bolt might not necessarily put on the nut. Assembling a Ford now took 90 minutes, though some other makes still cost ten times more because they were still made one at a time. Likewise, in the street railway industry, the people who ran each system, knowing that their city was unique, believed that their streetcars needed to be custom made. This precept always pushed costs up, however, and it was becoming increasingly distant from economic realities. As first-generation trolleys wore out, people in the transit industry began to realize that they were going to have to settle for some sort of standard replacement—manufactured in accord with Fordist principles—that could be sold at a price they could afford.

New cars were desperately needed in places where the same equipment had been in service since the turn of the century, but new cars would need to be cheap in every sense of the word: manufactured cheaply and cheap to operate, particularly in terms of platform costs but also with regard to the consumption of electricity, which had doubled with some of the larger streetcars. And if that meant a return to single-truckers, well so be it.

Streetcar systems in a number of smaller U.S. cities such as Tacoma and Bellingham, Washington, and El Paso and Galveston, Texas, were owned and managed by the Stone & Webster Corporation, a Boston holding company. It was just these sorts of systems that were most quickly put in a financial bind because of inflation and because of competition from motor vehicles. And the systems that most needed to cut expenses were usually the same ones that needed new equipment most urgently. Thus, Stone & Webster took the lead in developing a standardized streetcar that could be produced inexpensively, and it looked like this could only be a 4-wheeler. No matter how much of a throwback that might seem to be, such a streetcar would also have the virtue of minimizing power consumption. And then there were the platform costs: All indications were that it would be much easier to sell the idea of one-man operation to the public with a compact lightweight car than a large double-trucker.

Without a doubt, platform men had strong arguments in favor of keeping conductors' safety above all. Running a streetcar could be a demanding job, and a motorman who was distracted by passengers inside his car might not be able to pay full attention to what was going on outside. Deflecting concerns about safety was a primary concern of the engineer that Stone & Webster put in charge of the project, Charles O. Birney (1867–1939), who worked for its subsidiary Seattle Electric Company. Birney therefore devised a "dead-man" pedal that would cut the power and bring a car to a halt if the motorman ever lifted his foot without first setting the brakes. This was an important precaution and Birney's cars were dubbed "safety cars" as a reminder to the public that they would not go out of control if the motorman lost his concentration or somehow got disabled.

The first Birneys were delivered in the summer of 1916 in Seattle and nearby Bellingham. They were plain and simple, and very small—almost of Toonerville Trolley proportions. They were 28 feet long, with seats for only thirty passengers. They had no upholstery, not even rattan, just wooden benches. But they were priced at only $6,000, not a lot more than a big Packard sedan. Moreover, they *did* appear as if one man could handle them: "the number of accidents and the amount of damage," said one official after Birneys had been in operation for a time, "does not exceed to any perceptible degree those cars with two-man operation" (quoted in Duke 1957, 6–7). While they were double ended, most of them had only a door to the motorman's right, for both entering and exiting; some had back doors, but these were designated as "exit only" and passengers would have to push them open to get off.

Although Birneys were built by several firms, the majority—as might have been expected given its industry dominance—came from J. G. Brill in Philadelphia, along with Brill's subsidiary American Car Company in St. Louis. It happened that neither the Philadelphia Rapid Transit Company nor St. Louis Public Service, both very intensive operations, regarded Birneys as a viable option. But other big firms thought they saw financial salvation and scooped them up by the hundreds. The Detroit Department of Street Railways and the Eastern Massachusetts Street Railway both acquired more than 250 of them; Brooklyn bought 212, Public Service of New Jersey 200, Richmond 141, and the two systems in Los Angeles 140. Many street railways in towns of 50,000 and less acquired a modest fleet of Birneys with the intention of retiring all their other rolling stock.

Birneys were often called *frequent service* cars ("A Car in Sight at All Times" was the slogan), and the announced plan everyplace was to tighten up headways in order to compensate for the limited capacity. Transit officials expected, or at least hoped, that reduced platform costs would more than offset the added expense of operating more cars. But the consumers of new technologies form their own opinions and often affect their deployment. Another way to characterize a tradeoff is to say that a situation may get worse for some people as it gets better for others. A more favorable balance sheet for management, perhaps, and yet when Birneys were assigned to lines formerly served by roomier and sturdier cars, *patrons* rebelled. Even with reduced headways, rush-hour seating was inevitably at a premium.

Moreover, a single-trucker of less than 7 tons was bound to make for a rough ride, especially if track was no longer being kept up very well. At every joint between one length of rail and the next, every 42 feet, there tended to be a sag that induced galloping, the same old problem with single-truck cars. Standees sometimes had trouble keeping their balance. To add insult to injury, with only half as much horsepower as double-truck trolleys—streetcars usually had a motor on each axle—Birneys were slower getting from one stop to the next than riders had come to expect.

For these and many other reasons, the new cars immediately fell into disfavor except where they happened to have replaced equipment that was utterly decrepit. Birneys acquired all sorts of unflattering nicknames like "cornpoppers." As countless letters to the editor insisted, cornpoppers were just the latest attempt by greedy traction magnates to get richer at the expense of John Q. Public.

Birneys sold fairly well for the first 4 years, a total of 4,000 of them by 1920. After that, however, Brill and other car builders found customers for only 2,000 more—even though it was obvious that tens of thousands of old streetcars needed replacement. A double-truck version of the Birney had more capacity but was no more popular with patrons because it was still so slow and uncomfortable. And because assigning Birneys to busy lines in big cities tended to decrease patronage, they were eventually relegated to marginal shuttles.

Birneys remained in operation on into the 1930s in smaller cities such as Bakersfield and Fresno, Saginaw and Pontiac, Baton Rouge and Shreveport. In Fort Collins, Colorado, they lasted until 1951 (Figure 3.4). But in most places this size, it was clear even in the 1930s that the financial downturn would prove irreversible and the readiest way to economize would be to abandon rails in the street and switch to vehicles with rubber tires. Occasionally—as was the case in Shreveport—this would be a newly emerging hybrid, a bus with electric motors and trolley poles (two of them, side by side, to make the circuit), and called a trackless trolley. Most often it would just have a gasoline engine. Patrons would usually say that a new bus of any kind definitely beat an old cornpopper with wooden benches. And, for management, running gasoline buses on city streets whose maintenance was a public responsibility would become an increasingly tempting option. Many years later, there would be arguments about rail-borne transit being fundamentally superior to buses, no matter the size of the city. This argument always had to confront the hard economic reality that a bus did not cost as much as a trolley, that it needed only one man to run it, and that it did not require maintaining tracks and the streets around them.

Figure 3.4: The last Birney cars in regular service in the United States were operated by the Fort Collins Municipal Railway in Colorado, which had been under public ownership since World War I. On four occasions their continued use was affirmed in a referendum, but finally it proved too difficult to keep them repaired. They remained a subject of nostalgic longing, however—an article about these cars in *The Saturday Evening Post* was titled "Some of My Best Friends are Streetcars"—and in the 1980s a civic organization got one of them operational on a "heritage" line similar to the line in Lowell, Massachusetts. (Donald Duke photo, author's collection)

INTERNAL COMBUSTION

The prehistory of electric locomotion dates back to Robert Davidson and Charles Grafton Page in the 1840s and 1850s, a Scot and an American. The prehistory of the internal combustion engine dates to the work of Etienne Lenoir, a Frenchman, and two Germans, Nicholas Otto and Eugen Langen, at almost the same time, in the 1850s and 1860s. Their engines burned kerosene, which was distilled from coal, and they were intended for stationary use in manufacturing plants, not for conveyances. Nor were they designed to compress the fuel mixture in the combustion chamber as would be essential for the power to propel a vehicle. But these engines were produced by the hundreds and even thousands, and there was a link to the internal combustion engines used for the first motor vehicles. Among the employees at Otto and Langen's Gasmotorenfabrik Deutz were Wilhelm

Maybach and Gottlieb Daimler, who in 1885 each unveiled motor vehicles fueled by gasoline engines, as did another German whose name still resonates today, Karl Benz.

Maybach, Daimler, and Benz were born in 1847, 1834, and 1844, respectively. Charles Van Depoele, Leo Daft, and Frank Sprague were born in 1843, 1847, and 1857, respectively. All six of these men were of the same generation, their enthusiasm for human mobility stirred by the steam railroad. But their vision of a new technological age differed fundamentally. Motor vehicles would be self-contained and the ideal was that they would go anywhere there was a road. Trolley cars would go nowhere at all on their own; rather, they would require an elaborate infrastructure and a centralized power plant. Motor vehicles would no more be self-*sufficient* than trolleys, however. They would require a source of fuel, petroleum pumped from deep underground and then refined by a complex process. And ultimately they would require a different but likewise elaborate infrastructure that would be funded by taxes on the populace as a whole. But the first problems that motor-vehicle pioneers confronted were with the technology of the conveyances themselves, which initially rendered them unsuited to serving a general transportation function.

After Maybach devised a component of the gasoline engine that proved essential to reliable operation, the carburetor (the device that atomizes fuel before it is sucked into the combustion chambers), engineering advanced rapidly on many different fronts. But not manufacture. Even as trolleys had come to dominate the urban panorama in the United States at the turn of the twentieth century, motor vehicles were a curiosity, and that is what they remained until Americans like Ransom Olds and Henry Ford got involved in design and production a few years later. "The automobile is European by birth, American by adoption," begins a classic in technological history, which then continues

> The internal-combustion engine, upon which most automobile development has been based, is unmistakably of European origin, and both the idea and the technique of applying it to a highway vehicle were worked out in Europe. On the other hand, the transformation of the automobile from a luxury for the few to a convenience for the many was definitely an American achievement, and from it flowed economic consequences of almost incalculable magnitude (Rae 1965, 1).

Except that some of them got larger and more powerful by means of double trucks, with a motor on each of four axles rather than only two, the conveyance that Frank Sprague debuted in Richmond in 1888 was not changed in any basic way for more than 35 years. In contrast, the invention of Maybach, Daimler, and Benz was utterly transformed before World War I, as advanced European technology was adapted to low-cost mass production methods. During the 1920s the auto's "convenience for the many"—especially the convenience of being able to "stop and alight where one chooses," as Marcel Proust put it in *Remembrance of Things Past*—wrought a stunning social transformation. A new Model T cost less than $400, a second-hand ("used") car as little as $50. By 1930, one in twelve New Yorkers owned an automobile, one in eight in Boston and Chicago, one in four in Detroit and Seattle, one in three in Los Angeles. In 10 years, the number of automobiles registered in the United States nearly tripled, from 8.1 million to 23.1 million. This transformation would have a profound impact on the transit industry from without, but there was an impending transformation within the industry, a *defensive* measure, as well.

NO LONGER JUST *RAILWAYS*

As early as 1920, Grover Whalen, the municipal official whom Mayor Fiorello LaGuardia would designate as New York's "official greeter" (and later the president of the New York World's Fair Corporation), had published a book titled *Replacing Street Cars with Motor Buses*. Whalen originated the "ticker-tape parade," in which the open-top automobile played a primary role. He hated tracks in the street, and looked forward eagerly to the day when "the trolley can be relegated to the limbo of discarded things, along with the stage coach, the horsecar, and the cable car." At the same time, however, a preeminent figure in the automobile industry also spoke out on the subject of buses replacing trolleys. This was Charles F. Kettering (1876–1958), the engineer who, among many other accomplishments, had enhanced the auto's accessibility by devising a system to start the engine with an auxiliary electric motor rather than a hand crank. In 1920, Kettering became president and general manager of the General Motors Research Corporation, and the next year he was interviewed by the *Wall Street Journal*. Said Kettering, soon to be known simply as "the Boss":

> Development of the internal combustion engine is the outstanding achievement of the engineering world today. It marks the first time that power of any considerable size can be taken virtually anywhere. For this reason the bus system will win out over the trolley system. Even if buses cost as much as trolley cars, the tracks, overhead construction, etc. of the latter are so expensive that buses are bound to win out in the long run.

For those invested in urban transit, the runaway popularity of the automobile posed a dire competitive threat, that was obvious. But there was a more inclusive category, the "motor vehicle," which took in the still-rudimentary conveyances called motorbuses, or just buses. In their opinion of buses, street railway officials were divided. Some of them wanted nothing to do with anything that had a gasoline engine, for any such vehicle reminded them of jitneys and the wildcat competition that undermined profits during the war. But those who insisted that "real transit" *must* involve rails and electricity were a dwindling minority. A majority sensed that buses could actually be a weapon with which to counter the challenge of the automobile. "Flexibility" was their watchword: buses connecting with suburban developments beyond the older ones; new cross-town bus lines laid out in accord with shifts in travel patterns (not everyone who rode public transit needed or wanted to go downtown); greater maneuverability in the face of traffic congestion; and, more and more, outright substitution of buses for trolleys on lines with light patronage where the investment in tracks and wires no longer seemed warranted. Or maybe not *only* lines with light patronage—that was the big question when it came to substitution.

The Detroit Department of Street Railways, the Los Angeles Railway, Pittsburgh Railways, Cleveland Railways, and a handful of other companies had put motorbuses in operation as early as 1920, and so these companies were no longer "railways" exclusively. But there were only a few hundred buses being used for urban transit all told (many fewer than in intercity service), and their number did not increase substantially during a brief period when street railways prospered in the early part of the decade. Even in 1924, a survey by the *Electric Railway Journal* indicated that transit firms were operating only 1,200 buses, perhaps 2 percent of the entire urban transit roster nationwide. Transit was provided *entirely* by buses in only three towns: Everett, Washington; Newburgh, New York; and Bay City, Michigan.

Eight years later, however, the American Electric Railway Association changed its name to the American *Transit* Association. In so doing, it was simply adjusting to a new reality, as when it had earlier changed its name from the American Street Railway Association because a street railway was virtually by definition electric. By 1932, there were at least 20,000 urban transit conveyances that were not electric, and, streetcars were entirely *absent* from dozens of towns all across the country, including many of the Stone and Webster properties that had been a prime market for Birney cars. They were gone even from one of the larger cities in Texas, San Antonio, with a population of 230,000. When San Antonio "motorized" in 1932, streetcars nationwide still outnumbered transit buses by a ratio of 5 to 2. But the ratio was changing all the time, and so the question that loomed in the minds of everyone in the transit industry was this: Would trolleys remain predominant on the busiest lines, on systems in the biggest cities, or was Boss Kettering correct in predicting that buses were "bound to win out in the long run"—by which he clearly meant that buses would win out everywhere.

GASOLINE-ELECTRIC POWER

The buses that had debuted on New York City's Fifth Avenue in 1905 were an oddity in several ways. Because "influential residents" of like mind to Grover Whalen regarded streetcar tracks and conduits as a nuisance, the Fifth Avenue Coach Company was one of the few companies that went directly from omnibuses with horsepower to motorbuses. But those buses were hybrids. As the design and engineering of motor vehicles with gasoline engines normalized, they were usually equipped with a transmission that obliged the driver to use a low gear when getting started—that is, a gear which multiplied the torque of the engine—then to shift to an intermediate gear and then to a high gear that allowed the engine to turn over more slowly when the vehicle was up to speed. Shifting from one gear to another required depressing a clutch to disengage the driveline from the engine and then manipulating a lever whose action always seemed a little uncertain, even to a professional driver. For many people, "the most difficult thing about driving . . . was working the clutch and manually shifting from one gear to another" (Volti 2004, 75).

After many years, this process was rendered automatic in diesel-powered buses and, later, in automobiles, though in over-the-road trucks the driver still shifts manually; think of an 18-wheeler accelerating from a stop. But mechanical transmissions long remained a weak link in the engineering of heavy-duty motor vehicles. Even if a transmission could have been up to the test of accelerating and decelerating over and over in a ponderous double-decker, the jerky process of going through the gears would not have been pleasant for passengers. So, the Fifth Avenue buses were actually "gasoline-electrics," in which a gasoline engine turned an electric generator that supplied power to two 45-horsepower General Electric motors geared to the axles, just as in with a trolley car. There was no transmission and no clutch, but rather a "controller" handle as in a trolley, and getting up to speed was no jerkier than in a trolley.

The same kind of buses were later operated in a handful of other cities, most notably by the Chicago Motor Coach Company, which was under the same management as the Fifth Avenue company. In some sense these buses marked the realization of a dream going back to Robert Davidson and Charles Grafton Page—an electric conveyance that carried along its own source of current and was thus self-contained. On Fifth Avenue and the Chicago lakefront, service by gasoline-electrics was different from that of other public transit vehicles in other ways, too. It had a "first class" air, and everyone was guaranteed a seat. Photos showing smartly dressed riders atop the double-deckers bring to mind the clientele of a present-day tour bus, not men and women who look like they are on their way to or from work. A clientele like this was rare, gasoline-electric buses were rare, and indeed motorbuses of any kind remained a rarity.

MOTORBUS DESIGN AND ENGINEERING

At first, what was called a motorbus was usually just a truck chassis with hardly any springing, no shock absorbers, hard rubber tires, and a body that was almost always top-heavy and poorly suited for start-and-stop operation. Not until the early 1920s was there a type of bus that was "purpose built"— that is, with chassis and body built together from the ground up, rather than the one grafted on to the other. It was the design of two brothers from Oakland, California, Frank and William Fageol (pronounced Fadgl), and they called their vehicle a "safety coach" with the Birney safety car in mind, because the riding public needed reassurance whenever there was one-man operation, as was the case almost always with buses. Although they sold most of their safety coaches to interurban bus operations, a few found their way into local city service. In 1924, in an effort to raise money to buy a manufacturing plant in Kent, Ohio (near Cleveland), the Fageol brothers sold some stock in their firm to Brill and also to American Car & Foundry (ACF), the nation's foremost builder of railroad cars. After a complicated series of transactions, the Fageol brothers grew disenchanted with the venture and bowed out, leaving Brill as a subsidiary of ACF.

Though it was perhaps not evident at the time, Brill's streetcar business was in terminal decline: Whereas production had topped 1,500 cars in 1912 and totaled 6,800 between 1913 and 1924, only four times afterward would annual output be in three figures, and twice during the 1930s Brill would not sell one streetcar all year. For the Fageol brothers, on the other hand, the bus business was just taking off. They started a new enterprise of their own called Twin Coach Company, their primary intended market being urban transit. Like all other bus manufacturers, the Fageols had initially located their engines under an automotive-style hood out in front of the cowl and windshield. But the design of a new Twin Coach Model 40 (the capacity was forty passengers) mimicked a trolley car. The entrance was ahead of the front wheels, and so was the driver's seat, with the same maximum visibility enjoyed by a streetcar motorman. There were twin "pancake" engines (with cylinders opposed to each other instead of in a straight line or a V-shape) mounted under the floor behind the front wheels, each with its own drive-shaft. After the Twin Coach Model 40, city buses would begin to look less and less like an elongated, oversized automobile; more and more, this design much like a streetcar's would distinguish them from buses for interurban service, which were not intended to accommodate standees and were often called "parlor coaches."

First to buy Model 40s was the Chicago Surface Lines (CSL), the largest transit firm in the country, for service in newly developing suburbs. Like most other streetcar companies, the CSL had made no significant track extensions since before the war. Ultimately the Fageol brothers sold more than a thousand Model 40s to firms from coast to coast—from a subsidiary of the Brooklyn & Queens Transit to the Key System, headquartered in Oakland and serving San Francisco's East Bay. In 1930, Model 40s began running through a new tunnel connecting Detroit and Windsor, Canada, an intensive operation with 2-minute headways (Figure 3.5). For transit operators such as this Detroit & Canada Tunnel Corporation, Model 40s were the right vehicle at the right time—in large cities they were used to extend service where there were no tracks; smaller systems bought them to substitute entirely for streetcars. Though not much larger than a Birney car, they could make better time because they did not gallop like a Birney with its short wheelbase. And, at a time when buses had a reputation for a lack of durability, they were so sturdy that some of them remained in service for nearly 20 years, longer than most Birneys.

Meanwhile, the new firm called ACF–Brill came up with a bus design of its own, with the body and frame not just integrated but integral (now standard with autos, this is called "monocoque" construction), and many other companies got into bus manufacture as well. Some had names familiar from the automobile world such as Studebaker, Pierce Arrow, Dodge, Reo, and White; some were exotic, such as Fargo, FitzJohn, and Flxible. The Mack Truck Company, which had built sightseeing buses even before World War I, began manufacturing city buses, a Mack Model CL resembling a Twin Coach Model 40. The Ford Motor Company introduced a small bus called (somewhat misleadingly) the Model 70 that sold particularly well in Detroit, the Department of Street Railways buying nearly 1,000 of them.

Figure 3.5: A Twin Coach Model 40 is seen exiting the mile-long Detroit River tunnel in Windsor, as another enters. These buses were operated from 1930, when the tunnel was completed by the Detroit & Canada Tunnel Corporation, until after the end of World War II. The tunnel is now jointly owned by the municipalities of Detroit and Windsor.

Model 70s appeared in many other cities and towns as well, and the company adopted the slogan "The Sun Never Sets on a Ford Transit Bus." Yet the predominant name in the bus industry was not Ford, or Fageol or Twin Coach or Mack, but rather the Yellow Truck & Coach Manufacturing Company, a firm that had origins as a subsidiary of Fifth Avenue Coach and Chicago Motor Coach, building chassis for their buses after the world war cut off imports from Daimler and DeDion.

YELLOW COACH

The entrepreneur behind Yellow Coach (and also the Yellow Cab Company in both cities—yellow is the color most visible at a distance) was John D. Hertz (1879–1961), an Austrian-born one-time newspaper copyboy whose name resonates in the world of automobility to this day. Even while still operating buses in New York and Chicago, Hertz began to ponder a new idea—providing customers with the convenience of a taxicab, but enabling them to drive where they wanted and to "stop and alight" as they chose—and he sold Yellow Coach to the rapidly-expanding General Motors Corporation. GM formed a subsidiary with a plant in Pontiac, Michigan, formally named Yellow Truck & Coach Manufacturing, but in the transit industry it was always just "Yellow Coach." Yellow Coaches had pancake engines in the rear, rather than under the floor, which gave them a greater capacity for standees. By the mid-1930s, Yellow Coach had become to the bus industry what J. G. Brill had been to the streetcar industry 20 or 30 years before; time and again, Yellow Coaches were replacing Brill streetcars as transit firms elected to motorize.

As the Great Depression lingered, there was a general sense that street railways in towns that counted their population in less than six figures were going to disappear. It was not long, however, before street railways were being abandoned in cities much larger: besides San Antonio, there was Jacksonville, Florida, one of the most important industrial and commercial centers in the southeast; Tulsa, Oklahoma, known as "the oil capital of the world"; and Montgomery, Alabama, the one-time capital of the Confederate States of America, where a key episode of the civil rights movement would play out on a General Motors bus. By 1940, it was obvious to transit companies in even larger cities such as Seattle, Honolulu, and Houston that they could save a great deal of money by running buses on city streets whose maintenance was funded by gasoline taxes. Seattle and Honolulu abandoned part of their infrastructure, the tracks, but retained trolley wires for trackless trolleys. The Houston Electric Company's was a more usual response; it got rid of everything.

While buses still had limitations—none had the passenger capacity of a double-truck streetcar and few had the durability—the firms that manufactured them had an attractive inducement, generous terms of sale. Yellow Coach "directly financed the conversion from street cars to buses." (These were the words of the general counsel for General Motors in the course of later litigation, and they were what helped fuel rumors of a conspiracy that would take hold in the popular imagination.) But even with Yellow Coach thriving in the years just before World War II, even with an allied firm called National City Lines gaining control of small and medium-sized operations all across the country, nobody in the mass transit business would have conceded that Boss Kettering's prediction would *ever* apply to the trunk lines that transported tens of thousands of commuters back and forth to the suburbs every day. The assumption was that in cities larger than 250,000—in the 1930s, there were thirty-seven such cities in the United States—trolley cars were the *only* conveyances capable of providing adequate transit on primary routes.

MANHATTAN

For moving the masses rapidly, for "rapid transit," nothing could beat subways and elevateds that were isolated from ordinary traffic on "the surface." But there were only four American cities with rapid transit—New York, Chicago, Boston, and Philadelphia—and by the 1930s it seemed unlikely there would ever be any more. As for surface transit, management of some of the biggest systems was wavering in its commitment to trolley cars. For example, New Jersey's sprawling Public Service Coordinated Transport (PSCT) elected to get rid of most of its streetcars and streetcar tracks, though not electric power. PSCT switched to All Service Vehicles (ASVs) from Yellow Coach that could run either on internal combustion or electricity from overhead wires. Then, there was Manhattan, which was a special case because it also had its rapid transit lines. While these would always carry the bulk of the ridership, patronage on the surface lines was enormous, 500 million annually.

On Manhattan, more so than in any other city, it was essential to get riders aboard as quickly as possible—the expression was to "swallow" them—and it seemed that no vehicle could do this as efficiently as a double-truck streetcar. (The hybrid behemoths on Fifth Avenue had the same reputation as double-deckers in most U.S. cities—slow to load and unload.) If buses were to be substituted for the streetcars needed to serve Manhattan in rush hour, one transit engineer argued, it would require two and a half times as many vehicles and the result would be "an intolerable congestion" (quoted in Schrag 2000, 68–69). The partisans of buses had their own arguments, however. Because buses could move freely around obstacles, they would actually *reduce* congestion even if there had to be more of them in service. A streetcar might be enabled to get over fire hoses laying across the street with so-called hose jumpers, portable steel bridges placed on the track. But a fire truck that blocked the way? Any stalled vehicle? A streetcar was stuck. Buses loaded at curbside, while boarding a streetcar required a wait in the middle of the street. To be sure, the places where people waited were protected with signs and reflectors, sometimes barricaded with low walls or railings, and indeed they were called "safety zones." But there were oft-repeated tragedies of men, women, and children who thought they were waiting safely for a streetcar, who were not really safe at all. On top of that, nobody needed Grover Whalen to remind them how easy it was to trip over streetcar tracks, how slippery tracks could be, how much of a hazard to motorists and pedestrians, not to mention cyclists.

As it turned out, disputes about capacity and congestion, disputes about the technical merits of streetcars and buses, were only a proxy for a contest over political power, similar to what had gone on in Los Angeles during the 1920s—and indeed what is *often* at stake when technological choices are made. Municipal officials preferred bus lines that they could control themselves, and in the New York mayor's office it had become quite a tradition to oppose trolleys. For their part, those who owned one of the two major streetcar companies serving Manhattan, New York Railways, hoped that conversion to buses would enable them to raise the fare to 10 cents, as on Fifth Avenue, and to escape from paving assessments and the burden of maintaining conduits, which were far more expensive than trolley wires. So it happened that on many of the avenues and streets of Manhattan—Manhattan, of all places, where the horsecar had debuted a hundred years before—electric streetcars began to disappear: Second Avenue in 1933, Fourth Avenue and Madison in 1935, then cross-town lines. At the same time, however, the other major company with streetcar lines on Manhattan resisted. This company was called the Third Avenue Railway, and not only did it continue to operate lines on Third Avenue, but also on Tenth, Amsterdam, Broadway north of 42nd, and 42nd, 59th, and 125th street, as well as networks in the Bronx, New Rochelle, Mt. Vernon, and Yonkers. During the late 1930s, it built hundreds of new cars in its own shops, and for a few years it was almost as if New York was staging a showdown between two modes of surface transit, rubber tire on paved roadway versus flanged wheel on steel rail.

On New York Railways, a choice was posed between antiquated street-cars and modern buses. Many lines operated by the Third Avenue Railway had new streetcars, but their design largely adhered to old conventions. Company officials would have characterized these as "tried and true," but others remarked on the obvious, that "new" did not necessarily equate with "modern." (Ultimately, Mayor LaGuardia threatened to terminate the company's bus franchises in the Bronx if it did not abandon streetcars on Manhattan.) Right across the East River, however, the Brooklyn & Queens Transit Company would play a key role in a concerted effort to develop streetcars that nobody could doubt were modern, and that could definitely outmatch any bus, not only in advanced engineering, but also in comfort and styling, and in what Madison Avenue was beginning to call "rider appeal."

4

Rail or Rubber?

More than half of the trolleys in operation on American street railways in 1930 predated the World War. Some of these were carefully maintained, but, if a transit company had been having financial trouble, the maintenance of its rolling stock was almost certain to have been deferred, often for a long time. Even if new, most rolling stock *looked* old fashioned. Several manufacturers had tried to emulate "automotive styling," including Brill with what it called its "Master Unit" and the St. Louis Car Company with its "Rail Sedan." But none had been particularly successful in making the appeal to modernity that automobile manufacturers emphasized in their advertising, especially after General Motors initiated what became known as the annual model change, with different styling from one year to the next. In the face of competition from Detroit, it was becoming clear that the design and engineering of a new streetcar would need to be superior to any transit vehicle currently in operation. The capacity would have to match that of a traditional full-size double-trucker, about one hundred passengers. At the same time it would need to embody major improvements in comfort and performance, and, above all, to change the *image* of the streetcar.

The Birney had been a failure, but at least it was an instructive failure. The Birney had made it clear that a small streetcar lacking any semblance of "rider appeal"—not to mention the agility "to maintain its place in street traffic"—was not the answer. Rather, a new design needed to be almost the opposite of a Birney; or, rather, it needed to be the opposite in every regard but one. It would still need to be inexpensive, which meant that there would need to be a tight cap on manufacturing costs. With transit no longer very profitable even in major cities, loans for capital improvements were not easy to secure. Chicago's huge system went bankrupt in 1927. When Henry Huntington died that same year, he bequeathed a solvent streetcar system, and among big cities the Los Angeles Railway was not alone in that regard. Still, the street railway industry had long outlived its image as a cash cow, if not exactly the taint of men like Charles Yerkes—its past association with greedy traction magnates was what still impelled the popular resistance to fare increases. Its more pervasive image by the latter 1920s, however, was of an industry that the world was passing by. News of Huntington's death came on the same day as news of Charles Lindbergh's solo flight to Paris. Competition was heating up for the "Blue Riband," the mythic speed record for transatlantic liners, and high-end automobiles were becoming available with 12- and 16-cylinder engines. When patrons were polled about the streetcar's deficiencies, "too slow" was the response even more often than "too noisy."

THE PRESIDENTS' CONFERENCE COMMITTEE

Each year in the autumn the American Electric Railway Association (AERA) held a national convention where executives from operating firms and suppliers traded information and ideas. There were panels and papers on all sorts of technical matters, but throughout the 1920s the focus had been on two general problems, actually dire necessities: replacing thousands of worn-out streetcars and standardizing the specifications of the cars that were to replace them in order to hold down manufacturing costs. By 1929, there was also a dawning sense of another necessity, more subtle but even more pressing: bold steps to deal with an "image problem." It was with all this in mind that the AERA formed the Electric Railway Presidents' Conference Committee, charged with underwriting and developing a standardized streetcar that transit firms could afford, that would change popular perceptions, and that ultimately would rejuvenate an industry.

The name of the ERPCC was usually shortened to the Presidents' Conference Committee, or just the PCC. Altogether, its principals represented twenty-eight operating firms and nearly all of the largest: Brooklyn, Baltimore, Chicago, Kansas City, Los Angeles, Philadelphia, Pittsburgh, St. Louis, Washington, D.C. (the two big Manhattan companies were notably absent). Twenty-six firms that manufactured streetcars or streetcar components were also involved, including Brill and including General Electric and Westinghouse, which made most of the controllers and motors. The committee was chaired by Dr. Thomas Conway, Jr., a man still in his thirties (b. 1882), who was quite different from most others in the transit industry. First of all, he was a scholar: Conway had written his University of Pennsylvania doctoral dissertation on the modernization of electric railways and then become a professor of finance at Penn's Wharton School of Business. Second, he was a visionary: He had been an executive with several Midwestern interurban lines and was known as an enthusiast for lightweight high-tech equipment as the best way to counter automotive competition.

Conway combined realism and optimism. Far too often, he wrote in a leading trade journal, streetcars were "excessively noisy, obsolete in appearance and not calculated to stimulate riding or earn a profit." The competitor of the trolley, he continued

> ... is the private automobile and by the standards of the automotive industry the comparative progress of our industry in large part must be measured. Judged by that standard, the evolution of the urban railway car has failed to keep pace with its rival. The 1930 automobile is infinitely superior to the product of five years ago ...

Even if the transit industry had so far "failed to keep pace," Conway believed that it was not too late to catch up—*if only* it could develop attractive, well-designed rolling stock that would be within the financial means of companies that were posting modest profits at best. His first step was to appoint a chief engineer, Clarence Hirshfeld, who had been head of research at the Detroit Edison Company (where Henry Ford once worked) and was also an official of the National Electric Light Association. Like Conway, Hirshfeld held a postgraduate degree and had been a college teacher, but unlike Conway he had no direct experience with the design and operation of streetcars.

The idea of turning the engineering of the PCC project over to an outsider was not universally popular with the officials of an industry that almost always promoted from within. But Conway had a sense that hidebound company technicians and operating officials were a big part of the reason that street railways were in crisis. With a staff of thirty, Hirshfeld established headquarters at the Ninth Avenue Depot of the Brooklyn & Queens Transit Company and set in motion a program aimed at creating a thoroughly modern streetcar that manufacturers could sell at a profit even if it was priced substantially less than what they had charged previously for cars with the same capacity. The target was $15,000.

PCC OPERATION

Hirshfeld's staff devoted major attention to the design of the trucks, motors, and controls. The aim was to attain a rate of acceleration quicker than an automobile but sufficiently smooth that standees—even women in heels—would not be thrown off balance. The assumption in the industry had been that a trolley's rate of acceleration could be no faster than 3 miles per hour per second (that is, 10 seconds to reach 30 miles per hour). Conway and Hirshfeld knew that this assumption derived from the characteristics of the venerable General Electric K-type controller, with its lever that motormen moved back and forth to change resistance in the electrical circuit and thus accelerate or decelerate. There were only nine "points," or running speeds. Every time the motorman moved it from one point to the next, a streetcar would jerk, slightly but quite perceptibly to a standee. An automobile did not gain speed with an easy motion, either, because a driver had to go through the gears. But everyone in a motor vehicle was presumably seated. PCCs were to have a foot pedal with a hundred seamlessly-blended running speeds that would enable them to accelerate more smoothly than any other vehicle of any kind.

Conway and Hirshfeld saw PCCs as vehicles for technological progress, and that is how they envisioned their public reception. But of course progress is always a matter of individual perspective, and some people may perceive that they are losing out even as others gain. Besides improved technology, besides a revitalized image, PCCs also embodied another objective in the realm of economics, the reduction of platform costs. PCCs were to be single-enders, and implicit in the design was to be evidence that conductors could be eliminated once and for all, even from a large, high-capacity streetcar. Indeed, there would be neither a conductor nor a motorman, only an "operator." Passengers would enter and pay at the front, and the second set of doors in front of the rear truck would be for exiting only.

One-man operation was prohibited by law in some cities, including Chicago and San Francisco, and Conway and his committee knew that it would take some effort to reverse such laws. As it turned out, the opposition was occasionally intractable. Because of this, plans for a large fleet of PCCs in strongly pro-union San Francisco were scaled back drastically. The British Columbia Electric Company, which operated transit in Vancouver, went ahead and acquired PCCs on the assumption that a "full crew" ordinance could eventually be overturned, but, when this goal was not attained even after many years, BCER ordered buses from General Motors and took its PCCs out of service. One way or another, however, platform men usually lost out on this issue. In Washington, D.C., for example, streetcars "of superior design" were exempted from its two-man ordinance—"superior design" meant PCCs. Even in San Francisco the electorate eventually passed a measure enabling the Municipal Railway to use a single crewman "on all single-end cars built after January 1, 1939," which happened to be when the Municipal Railway ordered its first PCCs.

Motormen and former conductors with sufficient seniority to keep jobs as operators were often paid a little more, and in addition they could take a measure of comfort, literally, in an operating station designed in accord with what came to be called the science of ergonomics. Almost always before, a motorman had to stand up most of the time with his hand on the controller, and there was only a wooden stool when he wanted to rest. In a PCC, the operator sat in a padded seat on an elevated platform in front of a control panel with toggles for the bell, the lights, and the doors. At his feet on the left was a dead-man pedal, as with a Birney car, and with his right foot he controlled the power and brakes—the operating manual pointedly observing that this "method of operation leaves the operator's hands entirely free to issue transfers and make change for passengers." (Rules about "exact fare only" were not instituted until much later, as protection against armed robberies.)

PCC STYLING

Equally as important as reducing platform costs and quick, smooth acceleration was the overall "look" of the PCC. By the 1930s, industrial designers were eager to imbue vehicles with the characteristics of "streamlining" that mimicked the contours of the most modern airplanes and airships like Zeppelins. On railways and highways, the Burlington Railroad's "Zephyr" and the Chrysler "Airflow" would capture the popular imagination with their apparent capacity to "go like the wind" wherever they went. One could not imagine any such thing even with the newest trolleys; even a Rail Sedan from the St. Louis Car Company was staid and squarish, with rows of protruding rivet heads. And trolleys were noisy. They rattled over switches and crossings, wheel flanges shrieked when turning corners, and the gears groaned constantly. In a world of Zeppelins and Zephyrs, no conveyance would be more in need of a new image than the streetcar. Not only did it need to perform better in engineering terms, it also needed to put on a better "performance" as show business understands the word.

Conway and Hirshfeld understood this perfectly. PCC designs specified lightweight external panels fabricated from a steel–copper alloy called Cor-ten that could be welded together seamlessly. There were no visible rivets. Corners were rounded and the windshield tilted back. Effective measures were also taken to deaden sound and dampen vibration, notably by using a resilient sandwich of hard rubber between the wheels and the steel tires, and rubber cushioning many other places, even in the tracks in which the windows opened and closed. No streetcar had ever had a forced ventilation system as efficient as a PCC, or had better interior lighting. Yet, even with all their improvements, it was still possible to manufacture PCCs at substantially less than a large conventional streetcar of the past, a saving attributable to their high degree of standardization.

PCCs were built by three different manufacturers, the St. Louis Car Company, Canadian Car & Foundry (which took delivery of primered bodies from St. Louis because Canada imposed high import duties on finished products), and Pullman-Standard at its plant in Worcester, Massachusetts. Brill initially signed on to the PCC project, but grew disenchanted with Conway and Hirshfeld's penchant for secrecy and elected to develop a competitive streetcar called the "Brilliner" with input from Raymond Loewy, the famed industrial designer. Brill officials contended that it was just as well engineered as a PCC, and certainly there was a stylistic resemblance, yet the company found buyers for only forty Brilliners, the largest order going to Atlantic City, where the PCC committee had been formed. After having turned out more streetcars than any other manufacturer in history, Brill sold its last one to the Philadelphia Suburban Transportation Company in 1940 (Figure 4.1).

ACF-Brill Motors was not out of the business of making mass transit vehicles, not at all, for it would become the leading manufacturer of trackless trolleys, and in North America trackless trolleys would ultimately outnumber PCC cars. Which is certainly not to detract from the significance of Conway's achievement. The development of a stylish high-tech streetcar—in the depths of the Great Depression, it should be noted—was a brave initiative that fully deserves being called "one of the true high points in history of urban transit" (Cudahy 1995, 168). The PCC was what assured the survival of rail-borne transit on the streets of major American cities into the second half of the twentieth century. It is a classic of industrial design, still with the same air of modernity in 2006 as it had when it made its debut in 1936, and with many aspects of its engineering still reflected today in state-of-the-art light rail vehicles.

PCC ENGINEERING

PCCs embodied more than a hundred patented components, particularly involving the trucks, which were manufactured by a newcomer to the transit

Figure 4.1: Brill's answer to the PCC, a streamlined Brilliner, is seen here on private right-of-way on the western outskirts of Philadelphia. Brill had previously designed trolleys it dubbed Master Units whose "automotive styling" was carried over into Brilliners and then into rubber-tired transit vehicles. (Author's photo)

industry, the Clark Equipment Company of Battle Creek, Michigan (if Brill could not quite match the engineering of any one PCC component, it was the trucks). Patents were held by an entity called Transit Research Corporation, which collected royalties and used the proceeds to fine-tune the design of PCCs in various ways during the 15 years they were manufactured. For example, the brakes and doors were originally operated by compressed air, but most PCCs built after World War II were advertised as "all electric." Likewise, newer cars had a row of oblong "standee windows" above the regular windows that enabled riders who were not seated to see out and determine where they were and when they were approaching their stop (Figure 4.2).

Because PCCs were assembled on standardized but adjustable jigs, a certain amount of variation in the location of the doors, windows, and the overall length was possible, in accord with the preference of different operating companies but without added expense. Chicago (which continued with two-man operation) favored 50-footers, with doors front, center, and rear. Typically, however, a PCC was 46 feet long with two sets of doors, weighed 33,000 pounds, and had a capacity of fifty-five seated riders and about the

Figure 4.2: A postwar all-electric PCC with standee windows is seen here looping at the Calvert Street (now Duke Ellington) Bridge over Rock Creek. Washington, D.C.'s Capital Transit had one of the largest PCC fleets. (Author's photo)

same number of standees. PCCs had four 55-horsepower motors that transmitted their torque to the axles noiselessly by means of a propeller shaft and the same sort of spiral-bevel hypoid gears with which automobiles were equipped. They had dynamic brakes that turned the motors into generators on deceleration. And, not least important, there was the element of style. There were suggestions that the new "million dollar streetcar" be called a Cityliner or Streetliner—railroads such as the Union Pacific and Burlington were making headlines with their "streamliners." But the prosaic "PCC" stuck and became famous even though people could not remember what the initials stood for.

When Conway and Hirshfeld first released the engineering specifications for PCCs, the Brooklyn & Queens Transit Company (B&QT) requested bids for the production of a hundred cars. The Clark Equipment Company was awarded a contract for an experimental aluminum car (which served for 20 years but was never duplicated) and the St. Louis Car Company got a contract for ninety-nine more. The first of these to arrive in the fall of 1936 were put in service on a line that crossed the Brooklyn Bridge and looped on the Manhattan side. (The newspapers took due note of the irony: thoroughly modern streetcars would be dropping in from Brooklyn just as many of Manhattan's were disappearing.) When PCCs were put in service on the B&QT's long line to Surf and Stillwell on Coney Island, it posted a 33 percent increase in gross revenue. "Best thing that has ever happened to Brooklyn," wrote a respondent to a rider's poll, and that same feeling was shared in every other city.

The PCC reaffirmed "the prestige of the streetcar," remarked a trustee of Pittsburgh Railways, which ordered the first of hundreds of them in 1936, as did Chicago and Baltimore. The San Diego Electric Railway and the Los Angeles Railway placed early orders too. When the first PCCs were delivered in Los Angeles in March 1937, Mayor Frank Shaw proclaimed it to be "Transportation Week." Along with the president of the company, Shaw pulled the canvas cover from one of the cars and child actress Shirley Temple christened it with a bottle of champagne, just like it was an ocean liner. Philadelphia's first PCCs were paraded through the streets with a police escort. "Stops people in their tracks" said the president of Baltimore Transit. An ordinary citizen might not be able to afford a ride on the Burlington Zephyr or drive a Chrysler Airflow, but anyone with carfare could hop aboard a PCC right down at the corner. In Chicago, Frank Sprague lived long enough to see a PCC in operation on State Street.

PCC SALES AND RESALES

After Pearl Harbor, when the War Production Board restricted the acquisition of new equipment in order to conserve scarce materials such as rubber, would-be purchasers of PCCs had to qualify on the basis of a city's "strategic importance." Of course Pittsburgh qualified, as the nation's steelmaking capital. By 1943, Pittsburgh Railways had 401 PCCs, 40 percent of its entire roster, but accounting for much more of the total mileage because of their speed and reliability. And naturally Washington, D.C., qualified too. Between 1937 and 1946, Capital Transit placed nine orders, and within a few years it would become the first city with a streetcar roster consisting exclusively of one-man PCCs, 489 cars. Four cities had fleets even larger than that: 540 in Philadelphia, 666 in Pittsburgh, 688 in Chicago. Including cars purchased second-hand from Cincinnati, Cleveland, Kansas City, and Birmingham, Alabama, the Toronto Transit Commission eventually acquired 745 PCCs. Although small transit firms lacked the capacity to finance the purchase of PCCs, there was one notable exception. This was Johnstown Traction in Pennsylvania, which ordered seventeen of them in 1944 when its annual patronage was 25 million, but put them on the second-hand market in the late 1950s, when patronage was down to 6 million and dropping at the rate of 16 percent a year.

The loss of patronage in Johnstown was extreme but it was not much different in cities that were far larger. Transit ridership everywhere in America took a nosedive as automobile manufacturing came back to life in the latter 1940s and roads and highways began to undergo major improvements. The management in some cities hurried to get their PCCs up for resale while they were not yet fully depreciated. San Diego was first, in 1949, selling its PCCs to El Paso for the international line connecting with Ciudad Juarez; then Minneapolis and Birmingham sold their PCCs in 1953; Detroit and Dallas in 1956. The Birmingham Electric Company operated PCCs for only 7 years; some of the cars in Minneapolis operated for only 4 years. Louisville purchased a fleet of PCCs just after the war, then turned around without ever putting them into service and sold them to Cleveland, acquiring General Motors buses instead. Whether or not PCCs would have ultimately served the public better than buses in Louisville, and in many other places—that became a vexed question. Nobody disputed that in a city the size of Louisville it would be more economical in the short term for management to operate transit that did not need tracks in the street. But what about the quality of service? What about the long term?

Some 2,000 new all-electric PCCs were put into operation after the end of the war, and there were large orders from the largest cities: 210 went to the Philadelphia Transportation Company, 250 to the Toronto Transit Commission, and a whopping 600 to the Chicago Transit Authority. By 1951 there were more than 4,900 PCCs running in North American cities, operated by two dozen different transit firms. The PCC had been a success, no doubt about it, but it did not "save the streetcar industry" in the way that Thomas Conway had envisioned. Almost everywhere, transit patronage was slipping so fast by 1951 that only a handful of companies were solvent enough for major purchases of any kind of conveyance. (Not only were there more PCCs in operation in 1951 than ever before or after, but that was also the peak year for city *bus* operation, as well, the total being around 57,000.) By then, operating firms in cities much larger than Louisville were formulating plans for abandoning streetcars altogether within the next 10 or 15 years.

The twenty-five PCCs that the St. Louis Car Company delivered to the San Francisco Municipal Railway in 1951 and 1952 was the last order it ever filled (Figure 4.3). The same was true for Pullman Standard after it delivered fifty PCCs with "picture windows" to Boston. In the United States, there had been more than 27,000 trolleys in operation when wartime transit demand peaked in 1944 and 1945. Within 5 years, this number fell by half, to 13,225. Within 5 years more, the number was down to 5,300, nearly all of them PCCs. Along with Mexico City and Toronto, San Francisco

Figure 4.3: One of the last PCCs made in America is seen here at the west portal of the Twin Peaks Tunnel. San Francisco's hills and tunnels give its streetcars special appeal. In her memoir titled *I Know Why the Caged Bird Sings*, Maya Angelou tells about looking for a job during World War II, and how she especially fancied "the thought of sailing up and down the hills of San Francisco in a dark-blue uniform, with a money-changer on my belt." Though rejected repeatedly, she was determined: "I would be a conductorette and sling a full money changer from my belt. I would." She wrote that she never knew exactly why she was hired, only that "one day, which was tiresomely like all the others before it . . . the receptionist called me to her desk and shuffled a bundle of papers to me. They were job application forms. I was given blood tests, aptitude tests, physical coordination tests, and Rorschachs, then on a blissful day I was hired as the first Negro on the San Francisco streetcars." (Author's photo)

and Boston were among a handful of North American cities that remained committed to streetcars, at least on a few lines, and provided a market for used PCCs from other cities. Boston acquired twenty-five from Dallas after it abandoned its last streetcar line in 1956. San Francisco acquired sixty-six used PCCs from St. Louis Public Service, which abandoned its last streetcar line in 1966. The decade in between, from 1956 to 1966, marked finis for streetcars almost everywhere else, including Detroit, Chicago, Kansas City, Montreal, Baltimore, Los Angeles, and Washington, D.C. (see Table 4.1).

For a time, there was a lively second-hand market in other countries. Besides those that went to San Francisco, trolleys formerly operated on the streets of St. Louis also went to Tampico, Mexico. Ferrocarril General

Table 4.1
PCC cars in North American cities—years of operation

City	Years
Baltimore	1936–1963
Birmingham	1947–1953
Boston	1937–present
Brooklyn	1936–1956
Chicago	1937–1958
Cincinnati	1939–1951
Cleveland	1946–1952
Dallas	1945–1956
Detroit	1945–1956
El Paso	1950–1974
Illinois Terminal, St. Louis	1949–1958
Johnstown	1947–1961
Kansas City	1941–1957
Los Angeles Railway	1937–1963
Mexico City	1947–1975
Montreal	1944–1959
Toronto	1938–1990s
Minneapolis–St. Paul	1945–1953
Newark	1953–2004
Pacific Electric, Los Angeles	1940–1955
Philadelphia	1938–present
Pittsburgh	1936–1990s
St. Louis	1940–1966
San Diego	1936–1949
San Francisco	1948–present
Shaker Heights, Ohio	1948–1980s
Tampico, Mexico	1958–1970s
Washington	1937–1962
Vancouver	1938–1955

Urquiza acquired all thirty of the Pacific Electric's unusual double-ended PCCs for little more than it cost to ship them from Los Angeles to Buenos Aires. PCCs from the Los Angeles Railway went to Cairo, Egypt, which shared the nonstandard track gauge (3 feet, 6 inches) that L.A. had never changed after its cable car days. PCCs from Washington, D.C., went to Sarajevo, Yugoslavia, and Barcelona, Spain. The Toronto Transit Commission acquired hundreds of used PCCs. But a new life in another country was the best of fates. Much more typical was a trip to the scrapyard: Brooklyn's PCCs—the first to go into service—were scrapped by the New York City

Transit Authority in 1956. PCCs had been designed to last for 15 years, and they proved hardier than anticipated. But even if there had been interest in a replacement in a city like Brooklyn, there was no American manufacturer that would have been prepared to produce it.

THE TRACKLESS TROLLEY

In 1921, Charles Kettering would naturally have been excited by the prospect of conveyances with internal-combustion engines being used "virtually anywhere." But when he made that declaration, he also predicted that motor vehicles would soon be getting "100 miles to a gallon" and that pollution problems could readily be solved. Neither was true, and in fact air pollution would become a critical environmental concern as the number of motor vehicles in American cities multiplied after World War II. "Smog" first became infamous in Los Angeles and then in many other places. Internal combustion was nothing like a clean process, and especially not the combustion process of diesel engines, which became increasingly prevalent in buses. While gasoline buses greatly outnumbered diesels in 1945, the numbers were almost equal by the late 1960s, and by the 1970s the amount of diesel fuel being consumed by buses in urban transit was eight times greater than the amount of gasoline.

Diesel engines had marked advantages in terms of efficiency—the fuel was cheaper to begin with and the mileage was better—but "efficiency" is a factor that always entails tradeoffs. Any kind of bus had certain advantages over streetcars, of course, not least being that they were maneuverable. In a few cities there were stretches of private right of way—or, as it was called in New Orleans, "neutral ground"—where trolleys were separated from other vehicles. But where there had once been such rights-of-way, it was often paved to provide more lanes for automotive traffic. Passengers were nearly always compelled to board and alight in the middle of the street, and streetcar operators almost never had any way to get around an obstruction, not even a double-parked delivery truck. All an operator could do was ring his bell incessantly. Hence the appeal of an alternative type of conveyance that seemed to partake of the best of two worlds: it picked up current from overhead wires like a trolley car, *and* it could be steered like a bus—steered over to the curb to pick up passengers and steered around obstructions.

When the number of PCC cars operating in North America peaked in the early 1950s, there was a considerably larger number of mass transit vehicles of a type that was distinct from a streetcar and yet just the same in one key regard. Whatever the disadvantages of rail-borne conveyances in terms of burdening transit operators with expensive and inflexible infrastructure, there was no denying a key technological advantage of electricity. Electricity enabled every vehicle along an entire transit line to draw power from a central generating station via overhead wires, rather than necessitating that each one generate its *own* power by means of internal combustion. Along with the PCC, there had been another effort to develop an electric conveyance that would revitalize the transit industry. This involved a hybrid technology that still used overhead wires, as with trolley cars, but not tracks in the street. Instead, the conveyances had rubber tires and steering gear, and the operator maneuvered them with a steering wheel. In Great Britain they were called trolley buses; in the United States they were sometimes called trolley coaches, but most often trackless trolleys.

Even though the lineage of trackless trolleys reached as far back as that of the rail-borne streetcar, they did not attract sustained attention in the United States until about the same time that the PCC was being developed by Thomas Conway's committee. Nor were they ever the focus of an industry-sponsored R&D program like the PCC. While one manufacturer built four out of every five PCCs, the St. Louis Car Company, a large number of firms developed and manufactured trackless trolleys on their own. They ultimately proved as attractive to transit patrons as PCCs, and, if the sales tell the tale, considerably more attractive to many transit firms.

An excellent book about trackless trolleys is titled *Transit's Stepchild* (Sebree and Ward 1973). But the trackless trolley is better considered as the streetcar's half-brother, or, even better than that, its nonidentical twin. It was born at the same time. In 1879, when Ernst Werner von Siemens first devised a means of transmitting electric power to a moving rail-borne conveyance from a stationary dynamo, he was also experimenting with a conveyance not tied to a fixed guideway, which he called an *Elektromote*. Its brief public appearance in 1881 was the first of a long series of tentative steps in the direction of "rail-less electric traction" on both sides of the Atlantic. Regarding an early proposal for a line in Brookline, Massachusetts, one of the transit industry's primary trade journals commented:

> The trackless trolley, whatever may be one's judgment as to its commercial merits, is not a thing to be turned down offhand in these days of automobiles. It is an automobile system with a continuous source of energy, being thereby limited in its sphere of action, but relieved of the necessity of carrying a prime mover with it.

THE LAUREL CANYON LINE

One pioneer effort in the United States may be said to have a similar claim to fame as Leo Daft's 1885 operation in Baltimore—which was, in Frank Sprague's phrase, "the first regularly operated electric road in this country." This time, it was the first regularly operated trackless trolley line, and the location was the newly emerging motion-picture capital, Hollywood, California, 25 years later.

Charles Spencer Mann was of a sort never absent in southern California annals, a land developer. In 1910, Mann subdivided a residential tract at the top of one of the canyons in the Hollywood Hills, Laurel Canyon. He called it Bungalow Town. Passing by the mouth of the canyon were the tracks of the Los Angeles Pacific (LAP), the interurban trolley line that first linked L.A. with the beaches to its west and southwest, Santa Monica and Venice, Hermosa and Redondo. (The line was promoted by one Moses Sherman, and the construction was locally depicted as "Sherman's march to the sea.") Mann was in a similar position to those who had developed Angeleno Heights in the 1890s; he had assured prospective residents that he would provide transportation for the climb up the canyon from the LAP stop in West Hollywood, about a mile and a half. He had in mind gasoline buses, but the grade was tortuous and the conveyances he tried out were not up to the test. Besides the general problem, transmissions and clutches that were too frail, there was a problem specific to Mann's steep route: on upgrades, automotive radiators were notorious for boiling the water away and causing engines to overheat and (in a rather inapt expression) "freeze." Bungalow Town would need something reliable.

One day Mann saw an article about an experimental trackless trolley line in Zurich, Switzerland, and he got an idea about using a "continuous source of energy" for his Laurel Canyon buses. He enlisted LAP linemen to erect poles on either side of the road up the canyon, with cables spanning each pair. This was standard practice with streetcar lines. What was different here was that *twinned* trolley wires were attached to the spans—one for each half of the circuit. Mann bought electricity from the LAP, which had a substation at Sherman Junction in West Hollywood. For conveyances, he had the LAP mechanics strip two 16-passenger Oldsmobile buses of their gasoline engines and outfit each with a pair of Westinghouse streetcar motors, thereby relieving them of "the necessity of carrying a prime mover." On the roof were mounted a pair of spring-loaded poles with carbon "shoes" designed to bear up against the overhead wires. The shoes were unlike the wheels at the end of the trolley poles on streetcars, and paired poles had been seen with trolley cars only in the rare instances where the rails were not used for grounding, as in Cincinnati. But the design was different mainly in that the poles had swivels at the base so the shoes could stay in contact with the wires even as the vehicles swung off to one side or the other while negotiating the canyon road.

Mann kept his trackless trolleys going for nearly 5 years, charging a premium 10-cent fare. But one of the reasons why buses were not initially regarded as serious competition for trolleys was that they were not built to last very long, and by 1915 Mann's were worn out. Feeling flush from his successful real estate venture, he made some test runs with a big Stanley Steamer that was innocent of all the problems from which the Oldsmobiles had suffered at first (steamers had no transmission and no radiator), and eventually he bought two Stanleys and had the trolley wires taken down. Hence the parallels with Daft's Baltimore operation: Service eventually ended, but not until a period of continuous operation imparted to others a sense of technological possibilities.

EARLY VENTURES IN URBAN TRANSIT

In 1913, Edison's one-time collaborator Stephen D. Field organized the Electric Bus Company and built a prototype trackless trolley. Field was unable to stir up any interest and Electric Bus failed. But not long afterward the Atlas Truck Company of York, Pennsylvania, built seven "Trollibuses" in concert with General Electric and sold these to a transit company on Staten Island, New York. At around the same time, the prestigious Packard Motorcar Company sent a pair of trackless trolleys to Toronto for a tryout. Atlas had not been a name in the transit equipment industry, nor had Packard for that matter, but this was certainly not the case with the St. Louis Car Company, which built a "Trackless-Trollicar" prototype that got trials in Detroit and also across the river in Windsor, Ontario. Nor was it the case with J. G. Brill, which sent "Brill Railless" prototypes to Minneapolis and Los Angeles in 1922: 22 feet long with a capacity of fifty riders, seated and standing, and the same pair of 25-horsepower motors as a Birney, but almost twice as fast. Brill sold three of these to Baltimore's Union Railways & Electric Company for service on a 6-mile line to a new subdivision at Randallstown, and they operated for a decade, from 1922 to 1931. Trackless trolleys from the Brockway Corporation operated in Rochester and Albany, New York, for about the same limited period of time. But in Philadelphia the outcome was different: After a Brill Railless first appeared on Oregon Avenue in 1923, trackless trolleys (later renamed electric trolley buses, ETBs)

Demonstrating the flexibility of Rail-less Car Service in Philadelphia.

Supplement Your Railway Service with
Brill Railless Cars

The most important reason why Brill Electric Rail-
less Cars are recommended for supplementing existing
railway service is their comparatively low operating
cost. In districts where current is available, and there
is a possibility of ultimate railway extensions when
sufficient travel warrants them, Brill Rail-less Cars are
particularly adapted.

Present railway organizations are generally familiar
with this type of equipment, and therefore the maintenance problem is a simple one.

THE J. G. BRILL COMPANY
PHILADELPHIA, PA.
AMERICAN CAR CO. — G.C.KUHLMAN CAR CO. — WASON MAN'F'G CO
ST. LOUIS, MO.   CLEVELAND, OHIO   SPRINGFIELD, MASS.

Figure 4.4: The first standardized, production-model trackless trolley was the Brill Railless, which debuted on Philadelphia's Oregon Avenue in 1923, where this advertising photo was taken. Trackless trolleys ran continuously in Philadelphia for 80 years, from 1923 to 2003.

would remain a part of the urban panorama in Philadelphia until after the turn of the twenty-first century (Figure 4.4).

But aside from a few modest successes in the eastern United States—"Trollibus," "Trackless-Trollicar," "Brockway," "Railless"—the important developments with trackless trolleys were in Great Britain, where there were 400 of them in service by 1931. That was when the general manager of London United Tramways, alarmed by declining receipts, began substituting trolley buses on tram lines where the rails were in need of replacement. Under Lord Ashfield (who had once managed the Detroit Department of Street Railways and the vast system in northern New Jersey) the London Passenger Transport Board accelerated this conversion program. In the early

1920s there had been 2,600 tramcars in London. By 1940 there were less than 900 trams, but London had the world's largest trolley bus fleet, numbering 1,764.

At the same time, trackless trolleys gained ground slowly but surely in other parts of the United States. First to place orders, usually with Brill, were small operations that were in need of new equipment—often they had worn-out Birneys—or in some cases faced paving assessments they could not afford: Salt Lake City in 1928; Rockford, Illinois, in 1930; Knoxville and Memphis, Tennessee, in 1930 and 1931; Peoria, Illinois, and Providence, Rhode Island, in 1931; Topeka, Kansas, and Kenosha, Wisconsin, in 1932. By the mid-1930s trackless trolleys were showing up in bigger cities as well—New Orleans, Brooklyn, San Francisco, Chicago, Cincinnati, Cleveland—particularly as management in those cities warmed to the idea minimizing losses on lightly patronized ("thin") routes (Figure 4.5).

TECHNOLOGICAL ADVANTAGES

By 1940, there were more than 2,800 trackless trolleys in operation in cities from Honolulu to Winnipeg to Atlanta. For operating firms, there were two main advantages. Nobody presumed that a trackless trolley required a conductor, so one-man operation was not a political issue. Nor was there any longer a need for costly track maintenance and renewal. And yet, trackless trolleys could draw power from generating and distribution systems that were already in place for streetcar lines. When trackless trolley lines followed streets where there also were streetcars—or, more likely, where they had *replaced* streetcars—they entailed minimal investment in new infrastructure because the twin wires could be suspended from the same cables spanning the street and draw current from the same feeder that was already strung from one pole to the next. There were other advantages for operating firms and for patrons alike. Because trackless trolleys could be steered over to the curb even as the trolley poles remained in contact with the wires, they did not require safety zones. There were no inevitable delays when the roadway was blocked. Even though front-entrance one-man operation might slow boarding and lengthen dwell times, they made up for this with quick acceleration, a significant advantage whenever there were frequent stops. A survey conducted among transit riders on behalf of trackless trolley suppliers showed overwhelming approval for their "get up and go," their "ability to climb hills," their "freedom from fumes"—and especially their "silence," which was in marked contrast to any streetcar except a PCC.

Figure 4.5: The Brockway Corporation of Cortland, New York, tried to push the same advertising buttons as Brill — trackless trolleys for economizing on marginally profitable routes — but got contracts only from nearby Rochester and Albany. (*Street Railway Journal*)

The only sound when one passed was the hiss of the carbon shoes on the overhead wires.

When a transit company acquired trackless trolleys, mechanics who were accustomed to working on streetcars could adapt readily, whereas a switch to motorbuses required retraining in a whole different set of skills. Moreover, buses needed to be stored indoors (long after the disappearance of horsecars, storage sheds were still called "carbarns") in order to insure starting in subfreezing temperatures, whereas vehicles with electric power could be stored in open "yards." Internal-combustion engines required overhaul every 80,000 or 100,000 miles, whereas a trackless trolley could go 150,000 miles. For state and local taxation, trackless trolleys were usually classed the same way as streetcars, and rarely were they required to pay license fees or display license plates, as buses were.

Most transit firms that eliminated streetcars during the 1930s switched directly to gasoline buses, including those operated in such cities as Tacoma and Houston by Stone & Webster, which had developed the Birney car during World War I. In Great Britain it was more common for trams to give way to trolley buses, and this was also more common in Canada, where there eventually were trackless trolleys in a dozen cities. Among smaller cities in the United States, only Wilkes-Barre, Pennsylvania, and Covington, Kentucky, converted entirely from streetcars to trackless trolleys. Over the whole decade of the 1930s, the number of buses in transit service increased by nearly 16,000, the number of trackless trolleys by only a quarter that number. But trackless trolleys often figured in postwar modernization plans with big systems. As a result of gasoline rationing, the unavailability of new autos, and factories running at maximum capacity, transit patronage reached an all-time high near the end of the war. For the first time in many years, transit companies had lots of money to spend. They purchased 2,000 PCCs after 1945, but three times that many trackless trolleys, 2,000 of them going into service in just a 2-year period, 1949 and 1950. The best seller was ACF-Brill's TC-44, which could accommodate almost as many passengers as a PCC car but was made largely of aluminum and weighed only 18,250 pounds, 15,000 pounds less than a PCC (a postwar Cadillac sedan with tail fins weighed at least 6,000 pounds). As a rule of thumb, the less a transit vehicle weighed, the more economical it was to operate. For nearly a decade in the 1950s and 1960s, there were more trackless trolleys operating in North America than there were streetcars (see Table 4.2).

TRACKLESS TROLLEYS IN OPERATION

Several established streetcar manufacturers had diversified into trackless trolleys before 1930, but in the early years of the Great Depression transit patronage sagged and large orders for any kind of conveyance were rare. (The largest number of streetcars to change ownership went second-hand from San Antonio to New York's Third Avenue Line.) In 1933 American manufacturers sold not one streetcar, but ACF-Brill delivered twenty trackless trolleys for Columbus, seventeen for Dayton, and five for Providence. Truck manufacturers such as Kenworth, Leyland, and Mack also began making trackless trolleys, and so did companies that made gasoline buses, such as Twin Coach in Oakland, Marmon Herrington in Indianapolis, and Yellow Coach in Pontiac. Trackless trolleys would eventually operate on

**Table 4.2
Types of equipment operated by U.S. transit
companies, 1926–1970**

	Trolleys	Trackless Trolleys	Motorbuses
1926	62,857	–	14,400
1928	58,940	41	19,700
1930	55,150	173	21,300
1932	49,500	269	20,200
1934	43,700	441	22,200
1936	37,180	1,136	26,800
1938	31,400	2,032	28,500
1940	26,630	2,802	35,000
1942	27,230	3,385	46,000
1944	27,180	3,561	48,400
1946	24,740	3,916	52,450
1950	13,228	6,504	56,820
1955	5,300	6,149	52,400
1960	2,856	3,826	49,600
1965	1,549	1,453	49,600
1970	1,262	1,050	49,700

Source: American Transit Association

more than seventy North American transit systems. They would carry a major part of the load in San Francisco, Seattle, Chicago, and Boston, as well as in smaller cities such as Dayton, Providence, and Shreveport. The Shreveport company placed four orders with ACF-Brill during the 1930s, three more in the 1940s, and by 1947 its equipment roster was nearly 100 percent trackless trolleys. While trackless trolleys are long gone from Shreveport, the longtime manager of the Dayton Transit Company, William W. Owen (1901–1990), always marched to a different drummer, and, as a result of his influence, trackless trolleys operate in that midsize Ohio city to this day.

In many cities, trackless trolleys were an option considered but not chosen. One reason was that electrical suppliers demanded cash in advance, but motorbuses, like autos, were available on credit. In two California cities, Oakland and Pasadena, there was a buzz in the newspapers about the impending arrival of trackless trolleys and twin-wire overhead was actually in place. But when streetcars made their last run, they were replaced with buses from General Motors—which, as it happened, provided financial backing for an operating firm called National City Lines which had purchased the transit systems in both those cities. On the other hand, managers of firms that did acquire trackless trolleys were rarely disappointed. Almost never were they put on the second-hand market, as was so commonplace with PCCs. In some cities, PCCs ran for only a few years; trackless trolleys were often kept in service long beyond the time when they would have been fully depreciated. As suppliers went out of business or into other lines of manufacture, replacement parts had to be fabricated in company shops, and yet the total number of trackless trolleys in service in the United States and Canada did not substantially decline until the latter 1950s. There were trackless trolleys in Brooklyn for 30 years, in Columbus for 32, in New Orleans for 38, in Chicago for 43. But, when Chicago's last ACF-Brill TC-44s finally wore out in 1973, they were replaced with diesel buses.

One manufacturer continued to produce trackless trolleys (by the 1970s usually called ETBs)—Flyer Industries in Winnipeg, Manitoba—and Flyer found customers in Toronto, San Francisco, and Dayton. But trackless trolleys disappeared from nearly every other North American city, and, as that happened, urban transit seemed close to severing the allegiance it had sworn to electricity since the 1890s. It should be emphasized, however, that this was the situation only on "the surface"—electrified rapid transit became a growth industry in many parts of the world, and even on the surface it was true only in some countries. Today, there are nearly as many ETB systems worldwide as there are tramcar systems, several hundred of each. There are dozens in Russia and Ukraine, and there are ETBs even in China, where there are *no* trams. Nevertheless, in its devotion to the development of highway systems and to what is called motorization, China is presently following the same road that the United States followed for much of the twentieth century.

MOTORIZATION

By the second half of the twentieth century, surface mass transit in the United States was a declining industry—and the manufacture of mass transit conveyances was a *dying* industry. Just as the St. Louis Car Company took its last order for PCCs in 1951, the largest manufacturer of trackless trolleys, ACF-Brill, closed its books as well. Pullman Standard delivered its last trackless trolleys, the last of about 1,900, shortly after it closed out its books on PCCs. While one firm had dominated the PCC market, St. Louis, the production of nearly 10,000 trackless trolleys was much more diversified. Marmon-Herrington, an Indianapolis truck manufacturer

Figure 4.6: When the Bay Area Rapid Transit (BART) began running electric trains through its transbay tunnel, it was in essence reviving the Key System's service which connected San Francisco with the East Bay via the Bay Bridge beginning in 1939. Looking quite modern, here is the station at the end of the Key System line near the University of California just before the end in 1958, when tracks on the bridge were removed to make way for additional lanes of automotive traffic. (Author's photo)

(Col. Arthur W. S. Herrington helped design the jeep), turned out 1,543 of them. Twin Coach's total was 658. Mack Truck's was 290. Kenworth, Atlas, and Leyland, three other truck manufacturers, also made trackless trolleys. But for nearly all these companies they were a short-lived sideline, even as the demand for trucks grew exponentially with the decline of the railroads and the development of the interstate highway system.

Given the social, economic, and political factors favoring motorization in the United States, by the early 1950s it seemed that the PCC and trackless trolley could only forestall the inevitable. Not that there were not some highly regrettable moves, such as the abandonment by National City Lines of the electric railway linking San Francisco with the East Bay via the San Francisco-Oakland Bay Bridge—in order to create more lanes for motor vehicles (Figure 4.6). Parkways and "limited access" highways were being constructed in the vicinity of several big cities even before World War II, and afterward there were new "traffic engineers" whose mission was simply to devise means of expediting the movement of motor vehicles (Figure 4.7). Sometimes this involved paving over rights-of-way once reserved for electric railways, as on the Bay Bridge. More often, it was the

Figure 4.7: One of the early limited-access highway projects in the United States was this stretch of Ramona Blvd., east of downtown Los Angeles, which looks invitingly free of congestion in this 1937 photo. After World War II, passenger service on the Pacific Electric Railway line at the right would be abandoned, while Ramona Blvd. was transformed into the San Bernardino Freeway, now Interstate 10. (Courtesy Automobile Club of Southern California Archives)

conversion of two-way streets into one-way streets, which was usually a death-knell for streetcar and trackless trolley lines. If there had to be mass transit at all (and some of them seemed to be wondering about that), traffic engineers preferred buses because they could get around obstructions. Said one of them in Baltimore, "the only trouble with streetcars is that they are on the street" (quoted in Harwood 2003, viii). By 1975 there were streetcars left in just a handful of cities in the United States, plus those in Mexico City and Toronto. Mexico City and Toronto had trackless trolleys as well, and there were seven other Canadian cities with ETBs: Calgary, Edmonton, Hamilton, Kitchener, Saskatoon, Toronto, and Vancouver. With the five in the United States—Boston, Dayton, Philadelphia, San Francisco, and Seattle—that came to thirteen North American cities, when there had once been six times that many.

The number of trolley bus lines likewise dwindled in Great Britain and many parts of Europe, as did tram lines, but this was certainly not the case everywhere. In some countries the political context was different and there was not the same sort of overwhelming push for motorization. In the Netherlands, Belgium, Germany, Switzerland, Czechoslovakia, and all of eastern Europe, tramcar and trolley bus systems thrived throughout the second half of the twentieth century. Many tram systems were completely rebuilt after destruction by aerial bombardment during World War II; many continued to expand. By the early 1980s, Leningrad (St. Petersburg) was operating more than 2,000 tramcars built by a Russian firm, Yegerov, and today St. Petersburg also has 1,000 ETBs. Moscow has 1,700. In Moscow, Bucharest, Vienna, Budapest, and Prague, in each of these cities, there are thirty or forty tram lines, each with 800 to 1,000 cars.

Before World War II, there were nearly two dozen trolley systems this large or larger in the United States, but even as early as 1940 transit buses outnumbered streetcars nationwide, 35,000 to 26,600; almost every urban transit company was operating buses on part of its system, and more than half of them had nothing *but* buses. By the 1970s, the transit authority in northern New Jersey would have a fleet of 3,000 diesel buses, a number exceeded only by Transports en Commune de la Regione Parisienne with 4,000 and the London Passenger Transport Board with 6,000. The total eclipse of streetcars in parts of the world where it first flourished, the United States, Britain, and France, was the most notable turn of events in urban mass transit since electricity had eclipsed the horsecar and cable car at the end of the nineteenth century. In 1921, when Boss Kettering had predicted that the bus system would "win out over the trolley system"—in America if not eventually everywhere—not many people would have agreed, at least not that it would win out so decisively. Even if this was going to happen within only 15 or 20 years in cities in San Antonio's range—the fifty-six with a population between 100,000 to 250,000—what about the 24 ranging from 250,000 to 500,000? And the thirteen larger than that? Could it be possible in great cities with thousands of streetcars and passenger loads in the hundreds of millions?

It was indeed possible, even with the longest and most heavily traveled lines like Clark-Wentworth in Chicago, Germantown Avenue in Philadelphia, and Towson-Catonsville in Baltimore. And even with the most comprehensive systems: At the end of World War II, Pittsburgh Railways had operated more trolley lines than any other North American city; by the 1980s there was only a handful left—those running through South Hills Tunnel—and a publicist for the Port Authority of Allegheny County could brag about "erasing the derogatory reputation Pittsburgh endured as the largest trolley museum in America" (quoted in Kashin and Demoro 1986, 94).

As trolleys disappeared nearly everywhere in the United States, transit riders might be pleased by the novelty and newness of buses, but devotees would bid teary farewells. In *Dandelion Wine*, Ray Bradbury captured the mood:

> ... no matter how you look at it, a bus ain't a trolley. Don't make the same kind of noise. Don't have tracks or wires, don't throw sparks, don't pour sand on the tracks, don't have the same colors, don't have a bell, don't let down the step like a trolley does.

After an isolated line on the Queensborough Bridge was abandoned in 1957, there was not a single streetcar operating anywhere in New York state, where they had once run in two dozen different cities. In 1958 streetcars disappeared from Chicago, and a few years later from Washington, D.C. Not long after the last Washington PCCs were trucked to a museum, ground was broken for Metro, one of most ambitious and expensive construction projects in history. But Metro was largely designed to attract the patronage of suburban white-collar commuters. Transit on "the surface," the kind that had once provided the lifeblood of cities? Only buses.

DIESEL POWER

There are many reasons why transit patronage hit the skids in the United States after 1945—the auto's lure of independence, for one, and certainly the apparent convenience afforded by new urban freeways. And there are many reasons for the motorization of transit, some of which seem to lend credence to the widespread belief that corporate villains played an instrumental role. But not least was the almost continuous improvement in the design and engineering of buses, even as the economics of street railways disintegrated.

More and more of the buses that replaced streetcars after the war were a new breed named after a German engineer, Rudolph Diesel, who was born in Paris in 1858 and was just a year younger than Frank Sprague. Even as Sprague was fulfilling his contract to implement an electric streetcar system in Richmond, Diesel was working out his ideas for a new kind of internal combustion engine, in which the fuel was ignited by the heat produced by compression alone rather than atomizing it in a carburetor and igniting it with a sparkplug. Its great advantage was that it could burn "heavy oil," minimally refined, less expensive than gasoline, and containing more energy per unit of volume. In his 1893 patent application Diesel called it a "rational heat engine," and it was quite simple conceptually. But practical application was not a simple matter, and for 20 years before his death (apparently by suicide) in 1913 Diesel pursued developmental work in Germany, Russia, and France. Because they operated with such high compression ratios, diesels (after Rudolph Diesel's death, the name for "compression ignition" engines became generic—that is, with a lower-case "d") had to be exceptionally sturdy, and that meant they were heavy. At first, they were suited only to applications where weight was not crucial, for stationary power plants in factories and to some extent in maritime service. Because of their torque and fuel economy, diesels also began to find a niche in heavy-duty trucking—an American firm headed by C. Lyle Cummins was particularly active in developing diesels for trucks—and Daimler Benz in Germany and Citroen in France both produced automobiles with diesel engines. But as late as 1937 transit firms in the United States were operating only sixty-two diesel buses.

An important breakthrough had already been made at General Motors Research, however, where Kettering adopted as his "pet project" a diesel designed to operate on the so-called 2-stroke cycle. Though 2-strokes could be lighter and more powerful, before Kettering devised a means for "scavenging" the cylinders, such engines had been deficient in their capacity to expel exhaust gas and compress a fresh charge of fuel at the same time. Now, Kettering began to envision diesels "in a wider range of activities." Most notable were engines for the first streamlined trains such as the Burlington Zephyr. Kettering himself never pursued 2-stroke automotive applications seriously, but an engineer at the Yellow Coach factory, George Greene, was of a different mind. Formerly a motor-transport officer in the British army and an engineer for the London General Omnibus Company, Greene had his own pet project, development of a practical hydraulic transmission, which he believed would be the key to the success of diesel buses in urban transit.

Greene was right. Buses with GM 2-stroke engines and hydraulic transmissions proved their capabilities on the busy Eighth Avenue line in New York. After that, managers of transit systems almost everywhere could envision such buses in a role much larger that "feeders" to streetcar lines, and by 1940 there were more than ten times as many diesel buses in operation all across the country as there had been 2 years earlier. After World War II the bus that General Motors called the TDH (transit, diesel, hydraulic) commanded cityscapes from coast to coast (Figure 4.8). Ultimately GM's dominance in the production and sale of transit buses was so complete that it was sued by the U.S. Department of Justice and convicted of violating

Figure 4.8: The technical data-sheet on this bus reveals that it was built by in 1948 by the GMC Truck and Coach Division, Pontiac, Michigan; designated type TDH (transit, diesel, hydraulic transmission); and delivered to National City Lines of Chicago, then sent to Terre Haute, Indiana, and then in late 1954 to Montgomery, Alabama. There were countless GM TDHs, but none remotely as historic as the 2857 because of what Rosa Parks did when she was told to "move to the back of the bus." (From the Collections of The Henry Ford, copy and reuse restrictions apply http://www.TheHenryFord.org/copyright.html.)

antitrust laws by creating what is best described as "a sole-source supplier relationship with effectively captive consumers" (Jones 1985, 63). But GM was never convicted of a more sweeping charge, still given credibility by many people today, that it conspired to wreck viable electric railway systems. Actually, the mass transit industry in the United States might have failed altogether were it not for the financial relief provided by diesel buses.

That said, it must be added that dieselization also had a negative consequence about which there is no room for dispute. Men like Kettering and Greene brought the diesel engine and the diesel bus to a high degree of refinement, to be sure, but in one respect the engineering remained terribly deficient: Diesel buses polluted the air with their exhaust, and because they became notorious as "stink buggies" they were responsible for damaging the image of public transit much as it had been damaged by the excesses of men like Charles Yerkes. Buses were sometimes perceived as less an urban necessity than a public nuisance, and for a long time cash-strapped transit companies only rarely tried to make bus transit even marginally more attractive. The National City Lines successor in Montgomery, MATS

Table 4.3
Energy consumption by type of vehicle, 1945–1970

| | Kilowatt Hours (Millions) | | Gallons of Fuel (Thousands) | |
	Trolleys	Trackless Trolleys	Gasoline	Diesel
1945	4,547	520	510,000	11,800
1950	2,410	851	430,000	98,600
1955	910	720	276,300	172,600
1960	393	417	181,900	208,100
1965	218	181	124,200	248,400
1970	157	143	68,200	270,600

Source: American Transit Association

(Montgomery Area Transit System), long operated one of the most decrepit fleets in the country, and not until it finally upgraded with light-duty para-transit vans did it even put benches at bus stops, so people could sit down while they waited. Benches are still missing at bus stops in many big cities, as are rain shelters and signage with information on routes and schedules. There are few better examples than the diesel bus of a new technology yielding "economic benefits" (in terms of reduced expenses that were what enabled many transit firms to survive) while trading off "social costs" (in terms of environmental degradation and deteriorating quality of service).

For a while, Chicago was partial to propane, and other transit systems are now acquiring buses powered by "cleaner" sorts of combustion, with compressed natural gas (CNG) and liquified natural gas (LNG) recently coming into increasing use. Hydrogen and ethanol are current buzzwords. For several decades, however, a vehicle in urban transit was all but certain to be a diesel bus that lacked an effective means of controlling exhaust emissions. Between 1945 and 1970, the annual consumption of diesel fuel in urban transit grew from 11 million gallons to 270 million (see Table 4.3). Regional or metropolitan transit authorities could operate at a loss but still had to fight a desperate battle to keep costs under control, and when the chips are down there is no transit vehicle cheaper to operate than a diesel bus. The tradeoff, essential though it may have been, was probably the most drastic in the history of mass transit.

PCC TECHNOLOGY OVERSEAS

The predominance of the diesel bus in North America was echoed in most other parts of the world—indeed, on entire continents where there was (and is) almost no other mode of transit. But the diesel bus was not predominant in urban transit everywhere, not in some parts of western Europe or most parts of the Soviet Union; rather, the most common conveyances were in essence updated PCC cars. PCCs from the United States sometimes went second-hand to Europe, but more important was the persistence of PCC technology in second-generation trams manufactured in Europe under license from the Transit Research Corporation. Several hundred were made in Spain and in Italy, nearly 800 in Poland, more than 1,000 in Belgium (whence came financing for tram lines worldwide), many thousand in Russia and Hungary, and eventually some 20,000 in Czechoslovakia. Just as the St. Louis Car Company was delivering its last PCCs to the San Francisco Municipal Railway in 1951, Transit Research was licensing CKD Praha—known as Tatra Smichov—to manufacture PCCs for operation in the Czech capital, Prague. Tatra's first model was designated the T-1, and an improved T-2 followed that was intended for export. Then, beginning in 1960, Tatra began turning out its T-3, which would ultimately be manufactured in greater numbers than any streetcar in history. In 25 years, Tatra outshopped four times as many "PCC-type" trams as were made by American manufacturers from 1936 to 1951, and on dozens of systems in eastern Europe one can still see trams whose PCC lineage is quite apparent (Figure 4.9).

In the United States, a few original PCCs have been thoroughly rebuilt and they remain in everyday operation in Boston and Philadelphia and most notably in San Francisco, where they present a nostalgic panorama of living transit history, painted in liveries once familiar in many different cities all across the country. PCCs are preserved in several U.S. museums, from Baltimore to Perris in southern California, but the most interesting is in the National Tramway Museum in Crich, in Derbyshire, England, a tram that operated in the Hague, built to American designs in Belgium, and fitted with Czechoslovakian PCC motors that had originally been installed in a German tram. PCC engineering was also adapted to a different mode of transit, the sort that was less vulnerable to competition from motor vehicles because it was isolated from street traffic and was thus faster. When streetcars stopped operating in Chicago, its fleet of PCCs was rebuilt for service on the subway and elevated lines. Later, Pullman Standard delivered a fleet of cars designed to PCC specifications for the Cambridge–Dorchester line in Boston, and new cars designed to PCC specs also went to the Brooklyn elevated.

Mention of elevateds and subways, known as "heavy rail" or more commonly as rapid transit, carries this narrative into a realm that was foreign to all but about twenty cities in the world until after World War II, only

Figure 4.9: Many Tatra Smichov trams remained in operation at the turn of the twenty-first century; the scene here is Prague, near the Tatra factory. Though later models were articulated, with a truck in the middle, Tatra trams had virtually the same control and braking systems as had been devised by Thomas Conway's Presidents' Conference Committee in the early 1930s. (Author's photo)

four of them in America: New York, Boston, Chicago, and Philadelphia. New York's system was by far the largest and busiest anywhere: 1.8 billion riders a year in 1940, nearly four times as many as rode the surface lines. The underground system operated by the London Passenger Transport Board carried 600 million passengers annually; in Paris, the Chemin de Fer Metropolitaine recorded 800 million, and there were lesser numbers on less extensive undergrounds in Glasgow, Berlin, Budapest, Barcelona, Hamburg, Moscow, Madrid, and Vienna. Wuppertal had its monorail, and Liverpool had the one elevated railway ever constructed in Great Britain. Other cities around the world with rapid transit included Buenos Aires, Tokyo, Osaka, and Sidney.

The history of rapid transit in all its varieties—heavy rail, light rail, monorail, elevateds, subways, even busways—is deserving of a technography in its own right. Here, in the next chapter, the emphasis will be on one small aspect of that history, on the technologies that facilitated the initial development of rapid transit in American cities during the latter nineteenth century, and on the changes in political context that enabled a new wave of construction in the second half of the twentieth century.

5

Heavy and Light Rail

The phrase "rapid transit" was sometimes incorporated into the name of street railway companies and even bus lines; for many years, two brothers, J. T. and F. H. Asbury, operated a fleet of buses, mostly Macks and Twin Coaches, on lines connecting Pasadena, Hollywood, and many towns in the San Fernando Valley—and they called their company Asbury Rapid Transit. As in most other places, however, "rapid transit" came to represent a contradiction in terms, much like a new coinage of the 1920s, "rush hour." The one way to make urban transit *truly* rapid was to construct a right-of-way that was isolated from street traffic, either above it or below, and prior to World War II only in New York, Boston, Chicago, and Philadelphia could one actually get from place to place more rapidly than on city streets, known generically as "the surface." Their rapid transit lines became the envy of many other American cities with serious traffic congestion such as Los Angeles, Pittsburgh, Detroit, St. Louis, and Cleveland, all of which brought in consulting firms to develop plans for rapid transit, often doing so repeatedly. But it was never possible to secure financing while transit remained dependent, even partially, on private enterprise.

Beginning in the 1950s and 1960s rapid transit lines were constructed in North American cities where there had been none before, but only in Toronto, San Francisco, and Washington, D.C., did the new systems even come close to the scope of New York's, or even the rapid transit in

Boston, Chicago, and Philadelphia. The new subways in Atlanta, Baltimore, Miami, or Los Angeles could be tucked into one small corner of the New York system, which was built over a period of 40 years, from 1900 to 1940, mostly with local funding: 722 miles of track, 6,400 cars, 469 stations (Hood 1993, 12). In the intensity of its transit operations, New York had been in a league of its own ever since John Stephenson began building omnibuses for Abram Brower in the 1830s. Thirty years later, when New York had become the largest city in the country, there were twenty-nine companies operating omnibuses and fourteen more with horsecars, together carrying 100 million passengers a year. There also were dozens of ferries across the Hudson to New Jersey, across upper New York Bay to Staten Island, and across the East River to Brooklyn, which had an extensive transit system of its own. Yet 85 percent of New York's residents still lived within 2 miles of Borough Hall on Manhattan. The Third Avenue Railway, the principal horsecar line, was carrying the same number of passengers year after year, 20 million—that is, it was operating at maximum capacity. As one of the newspapers put it with some understatement, "the means of going from one part of the city to the other" were "badly contrived." Said another: "We can travel from New York half-way to Philadelphia in less time than the length of Broadway" (quoted in Bobrick 1981, 171).

Beyond 42nd, streets were ungraded and unpaved, and seemed likely to remain in that state until transit was improved. After 1863, London's Metropolitan Railway provided an example of one possible means of doing so—the Metropolitan comprised 3.5 miles of steam railway, partly in tunnels, partly in open cuts, that connected the Great Western's line at Paddington Station with the London & North Eastern's at Euston, the Great Northern's at King's Cross, and the Midland's at St. Pancras. But the Metropolitan did not dispel longstanding misgivings about steam railways in urban transit or newer concerns about underground railways: Would they devalue real estate? Was it not unhealthy to operate steam locomotives (even fueled by relatively smokeless coke) without adequate ventilation? Arguments that were cast in terms of technological alternatives often masked political motives. Shortly after the London Metropolitan opened, a proposal for a similar system in New York ran afoul of the entrenched system of political patronage. The power of New York's political boss, William Marcy Tweed, hinged directly on the monies paid to him by the companies that operated horsecars and omnibuses, and he was a staunch guardian of their profits.

Proponents of New York's Metropolitan could cite many advantages of a below-ground line. A passenger "would not be obliged to go into the middle of the street to take a car." Nor would a car "be obliged to wait for a lazy or obstinate truckman." On the other hand, it was easy to incite fears about recreating the "suffocating atmosphere" of London's Metropolitan, which was plenty smoky in spite of its coke-burning locomotives. And much was made of the problems that would be engendered by the total disruption of street traffic during construction, as well as the disturbance to underground utilities. Largely because of these two concerns, rapid transit on an elevated plane got serious consideration before subways.

THE AERIAL RAILWAY

At the same time that plans for the New York Metropolitan were being aired and ultimately defeated, a proposal for a "railway on stilts" was set forth by Charles T. Harvey (1829–1913), an engineer who had gained repute for his role in construction of the Lake Superior Ship Canal (also known as the Soo Locks), which opened the rich ore fields of the Mesabi Range to Great Lakes shipping. Boss Tweed, ever protective of the street-railway interests, fought against the proposal, but Harvey had a powerful political supporter of his own, Erastus Corning. Corning had been president of the company that built the Soo Locks, and was also the man behind the sweeping railroad consolidation (including the New York & Harlem) that resulted in the New York Central System, the great trunk line connecting New York and Chicago. Together, Corning and Harvey envisioned a 15-mile line from the Battery via Greenwich Street and Ninth Avenue to Yonkers, and the first segment of the West Side & Yonkers Patent Railway opened in July 1867.

Transportation technologies have two main components, conveyances and infrastructure, and no mode of transit can be developed satisfactorily without equal attention to both. With motor vehicles, the lag was in infrastructure; there were reasonably dependable cars before there was anything but a rudimentary highway system that would become impassable in bad weather. But the infrastructure is where Harvey perceived his main challenge, in the adaptation of its design from the technology of iron bridge building. Actually, his true novelty was in his conveyances.

Even before Andrew Hallidie in San Francisco, Harvey had devised a means of impelling vehicles by means of cables, using a "traveler" attached to the cable every 150 feet that would be engaged through a rotating arm on the car. A few years later, the limitations of a different sort of device that was sometimes called a "traveler" (that is, a troller) were what proved to be the undoing of Leo Daft in his efforts to develop a practical electric streetcar. A workable spring-loaded trolley pole with a wheel that bore up against an overhead wire was one key to Frank Sprague's success in Richmond.

Likewise, a grip that would take and release a cable gradually was the key to Andrew Hallidie's success in San Francisco. But Harvey never managed to make such a device work satisfactorily; it was either go or no-go, and that annoying jerkiness was part of his downfall. Even though he managed to secure funding for construction of his elevated structure as far north as Thirtieth Street, where Corning's railroad terminated, the shortcomings of his system of cable propulsion remained unresolved even as he was beset with financial as well as technical difficulties. After he was driven into bankruptcy in 1869, his patent railway languished. It was a story that would often be repeated in the 1880s with pioneering street railway ventures.

THE MANHATTAN ELEVATED

Eventually, after the West Side & Yonkers was been sold at auction, it was reorganized as the New York Elevated Railway, and it was from that point in October 1870 that the history of rapid transit in New York is formally dated—indeed, rapid transit anywhere. A few years later, the New York Elevated was merged with another failed elevated railway, that one above Sixth Avenue. This line had been the dream of Dr. Rufus Gilbert, a surgeon who had supervised the U.S. Army hospitals during the Civil War and envisioned rapid transit as an aspect of public health, a means of enabling the poor to escape the slums. What emerged from the consolidation of Harvey's and Gilbert's projects was a firm called the Manhattan Elevated Railway. With sufficient financing to keep 6,000 workmen on the job all at once, construction proceeded on several fronts and by 1880 elevated lines stretching from the Battery to the Harlem River were in full operation not only above Ninth Avenue, but also Second and Third, while the line begun by Gilbert above Sixth Avenue reached Central Park. The system now proven, an elevated railway was rushed to completion in Brooklyn in 1885, and within a few years in Chicago, Boston, and Philadelphia, as well, but none of these ever topped Manhattan's system in extent or patronage, or its impact on the popular mind (Figure 5.1).

Motive power for the Manhattan Elevated was provided not by cables, nor by electricity (whose practicality in urban transit was not to be conclusively demonstrated until 1888), but rather by steam locomotives that could operate with equal facility in either direction. These were the patented design of Matthias Nace Forney (1835–1908), author of the standard 1874 text, *Catechism of the Locomotive* and one of the founders of the American Society of Mechanical Engineers. At first, the public was fascinated with the sheer novelty of the elevated—especially with the "grand serpentine

Figure 5.1: This cut from the *Harper's Weekly* issue of September 7, 1878, contrasts the Manhattan Elevated Railway with the congestion below, on "the surface." Here, at Franklin Square, the tracks parted rather than running above the center of the street. Clearly this was a more rapid form of transit, but one can see why the elevated was a disturbing presence. Steam provided motive power for 32 years, from 1871 until 1903.

curve" on the Ninth Avenue line at 110th Street on the upper west side—and with Forney's "spunky little devils" that propelled the trains, 325 of them eventually. The "El" facilitated the development of many new residential neighborhoods on Manhattan and later in Brooklyn and the Bronx. In time it became clear, however, that this technological system entailed drastic tradeoffs. The structure left long stretches of New York's main thoroughfares in perpetual gloom. There were unending objections to smoke and soot—and especially to the noise, the "furious din that bursts upon the senses," as the novelist William Dean Howells wrote in dismay. Even though the owners of the Manhattan Elevated considered electrification in the 1880s and 1890s with "tractors" on the order of those that Leo Daft used with his early streetcar lines in Baltimore and Los Angeles, they remained loyal to Forney's steam locomotives until past the turn of the twentieth century. Steam actually outlasted Manhattan's cable cars. By the time electricity was finally substituted, Manhattan's first subway, the Interborough Rapid Transit, was almost complete and already there was talk of getting rid of the elevateds.

THE BEACH PNEUMATIC RAILWAY

As with Charles Harvey's "patent railway" and the Manhattan Elevated, the New York subway had a whimsical prehistory, this time involving Alfred Ely Beach (1826–1896), the youthful editor of a technical journal called *Scientific American*. Even before the Civil War, Beach had a plan:

> To tunnel Broadway through the whole length, with openings and stairways at every corner. This subterranean passage is to be laid down with double track, with a road for foot passengers on either side—the whole to be brilliantly lighted with gas. The cars, which are to be drawn by horses, will stop ten seconds at every corner—thus performing the trip up and down, including stops, in about an hour (quoted in Bobrick 1981, 172).

Beach did not foresee everything correctly, not "the road for foot passengers" or the horse power, but he had the general idea about a "subterranean passage" under Broadway, and he knew how to invoke positive imagery associated with the Underground Railroad, by means of which fugitive slaves were being spirited to freedom. New Yorkers, that is, would be freed from enslavement to the trolley barons. There was still the problem of impelling the cars—horses were proving increasingly unsatisfactory with street railways—and by the 1860s Beach had become an enthusiast for "pneumatic" power, the force derived by increasing or decreasing the volume of air in a given space. Actually, the idea was not so far-fetched; the British post office had already contracted for a network of pneumatic tubes to speed the distribution of mail in London, and this provided a model for systems installed in large department stores to send payment for purchases to a central cashier. The heart of the system was the Roots Patent Force Rotary Blower, recently developed by the P. H. and F. M. Roots Company of Connersville, Indiana, for ventilating mines and for other purposes yet to be defined. Beach's plan was to "blow" the car to one end of the line, then, when the wheels tripped a telegraph wire, an engineer stationed in the control room would reverse the blower and it would "suck" the car back to the starting point. In 1868 the state legislature granted Beach permission to construct pneumatic dispatch tubes under Broadway for transmitting letters and packages; he kept his real aim, "transmitting" passengers, a secret because he knew that Boss Tweed would oppose him, as he had opposed Harvey. But ultimately Beach's problems were like Harvey's: more technical and financial than political. Speed could not be controlled, nor could more than one car be sent through the tube at one time, nor was there any simple way of getting people on and off without some kind of an airlock. Nor was Beach able to raise money to try and work out his problems after the stock market crashed in 1873. His pneumatic railway languished, as had Harvey's railroad on stilts.

THE INTERBOROUGH RAPID TRANSIT

Twenty years later, in the face of growing dissatisfaction with the elevated railways—or, rather with the "unwanted results"—New York's Rapid Transit Commission mandated that any new lines be routed underground, but with some mode of propulsion other than steam, and definitely not compressed air. Planning for New York's first subway, the Interborough Rapid Transit Company (IRT), was put in the hands of William Barclay Parsons (1859–1932), a founder of one of the great international engineering firms of the twentieth century, Parsons Brinkerhoff (which would later be engaged in the construction of rapid transit lines from Atlanta and San Francisco to New Delhi and Cairo). After inspecting the tube under the Thames in London, Parsons felt certain that the IRT would require the same system of power distribution, an electrified third rail. Sizing up the challenge on Manhattan, he could see that part of the line between City Hall and 145th Street would require tunneling, as in London, but for the rest he would be able to use the technique that became known as cut and cover—much less expensive than tunneling, though the tradeoff was dealing with "a maze of electric cables, telephone lines, telegraph wires, water pipes, steam mains, and sewers that would have to be removed and rebuilt elsewhere" (Hood 1993, 84).

The long-range plan for the IRT was for the line to run the length of Manhattan and into the Bronx on one end and across the Brooklyn Bridge on the other (both Brooklyn and the Bronx became New York "boroughs" in 1898). After Parsons turned his proposal over to municipal officials in 1895, however, it took so long for it "to emerge from a labyrinth of mayoral, aldermanic, and judicial oversight" that subways were completed in Glasgow and Budapest in the interim and New York was even anticipated in Boston, where a subway began operation in 1897, 6 years before the IRT. The Boston project was of a different magnitude from the IRT, because the intention was simply to get trolleys underground in the part of town where traffic congestion was worst. But its existence ultimately proved instrumental in the survival of several of Boston's trolley lines into the twenty-first century, more than 50 years after streetcars disappeared from New York.

THE TREMONT STREET SUBWAY

Under the legislation enabling Henry M. Whitney to establish the West End Railroad, he was also authorized "to locate, construct, and maintain one or more tunnels between convenient points . . . in one or more directions under the squares, streets, ways, and places." In 1871, the annual patronage of Boston's streetcars had been 34 million; by 1881 that figure had doubled, and by 1891 it had more than doubled again, to 136 million. In order to reach the Boston Common, most of the West End's 2,600 streetcars had to be funneled along one street, Tremont, where congestion was, if anything, worse than on Broadway in New York. The initial proposal for alleviating this problem was an elevated railway on the same order as New York's. By the 1890s, however, elevateds had fallen into widespread disrepute and all it took to put that idea to rest in Boston was a doctored newspaper photo showing an elevated structure shrouding Tremont Street as a steam locomotive rumbled within inches of the steeple of Park Street Church. After that, Whitney turned his attention to a subway, even though underground railways were still so foreign to America that they were called "the European transit system." (Recall that horsecar lines had been called "American railways" in Europe as late as the 1870s.)

The Boston project, headed by chief engineer Howard A. Carson, called for a cut-and-cover tunnel under Tremont from the corner of Pleasant Street to just beyond Haymarket Square. The first segment, opened in 1897, reached as far as Park Street, and eventually every streetcar coming in from the West End dove underground, with downtown stops at Boylston and Park, and a loop at North Station. Boston's rather modest streetcar subway cannot be counted as an engineering milestone on the order of the deep-bore tube under the Thames or New York's 4-track IRT. But it eventually engendered more than its share of folklore, perhaps most notably the Kingston Trio's lament about Charlie, the man who could never return:

> Let me tell you the story
> Of a man named Charlie
> On a tragic and fateful day
> He put ten cents in his pocket
> Kissed his wife and family
> Went to ride on the MTA . . .
> Did he ever return,
> No he never returned

And his fate is still unlearn'd
He may ride forever
'neath the streets of Boston
He's the man who never returned.

Charlie's lament was owing to a fare-collection system instituted on Boston streetcars during the 1940s: Charlie was doomed to "ride forever" because passengers were required to pay another fare upon *exiting*—the amount varying according to how far they had traveled and whether they had transferred—and he did not have enough change. Had the Tremont Street subway been rapid transit in the full sense of the term—designed along the lines of the London tube or the IRT in New York—he would have paid his full fare before boarding, after purchasing a token and passing through a turnstile. But outside of the subway the Boston streetcar lines were not separated from ordinary traffic, and passengers had to board from street-level safety zones. A prepayment system with turnstiles would have been impossible to implement. Moreover, in the interest of speeding everything up on the IRT (and eventually on rapid transit lines everywhere, and even some light rail lines), station platforms were built at the same level as the floors of the cars, so riders could step directly aboard, with perhaps only a warning sign to "mind the gap." And there was one other distinction from what came to be known as "true" rapid transit. In Boston, if two of three streetcars were coupled together—as they often were—there still had to be a crew in each car, a critical consideration as platform costs became an increasing concern. For alleviating that problem, Frank Sprague would step to the fore once again, this time in Chicago, not long after streetcars first dove underground in Boston.

MULTIPLE UNIT CONTROL

The only elevated railway in Great Britain was in Liverpool, and elevated railways were nonexistent in Europe, where there was popular opposition even to the visual clutter of trolley wires. Many elevateds in the United States, on Manhattan and elsewhere, were eventually deemed a public nuisance and torn down. There are still elevated railway lines in the Bronx and Brooklyn, and even in Philadelphia and Boston, but Chicago is the one city where they remain integral to the downtown cityscape; indeed, the Loop is as much a part of the city's image as cable cars and trolleys are part of San Francisco's. The Loop was originally called the Union Loop and it was designed to unify the operation of elevateds coming into downtown Chicago from three different directions. Two of the lines were like New York's, with steam locomotives pulling trains that were necessarily short because close headways made rapid acceleration so important and longer trains would have been too slow. The third line, the Metropolitan West Side Elevated, had a 600-volt third rail mounted outside the running rails, and it used "motor cars" from General Electric that had a projecting "shoe" to contact the third rail. This was the first electrified rapid transit line in the United States, and, except for London, in the world. As with elevated lines that used steam power, however, the length of trains was still restricted by the inability of the motor cars to provide sufficient tractive power for speedy acceleration.

In 1897, one of the other elevated railways in Chicago, the South Side Elevated, was reorganized with the aim of converting from steam to electric motor cars like the Metropolitan's. General Electric, Westinghouse, and Siemens were invited to bid on the project. But the South Side's electrical consultant had worked for Sprague in Richmond and invited him to bid along with the others, knowing that he had something different in mind. Sprague had recently gone into business manufacturing the electrical motors and controls for elevators in the tall steel-framed buildings that were being erected in Chicago and New York. In New York, he had devised a system for controlling all the elevators in the Postal Telegraph Building from a single master switch, and he realized that he might apply a similar arrangement to electric trains. Each car would have its own motors and its own controls, but it would be possible to run the entire train from one set of controls in the lead car. The length of a train would no longer be limited by the capacity of one locomotive—whether steam or electric—to accelerate quickly with trailers in tow. Rather, the length would be limited only by the length of the stations: The "multiple unit control" (MU) system would amount to the ultimate minimization of platform costs.

Much as John Stephenson had thought of a way to design streetcars along different lines than steam railroad cars in the 1850s, lighter and more nimble, Sprague aimed to change an *operating* practice inherited from railroading. Instead of there being a steam locomotive or an electric locomotive car, every car would be motorized, and the electrical controller in each one would have what he called a "pilot motor" that would work in unison with the controller that was actually in the hand of the operator. By the late summer of 1898, Sprague had fulfilled his contract by equipping 120 of the South Side's cars with motors and MU controls. The advantages were manifold. Because traction was available from every car, the length of trains could be readily adjusted to meet traffic demands at various times of day: there could be just one of two cars at midday, then, during rush hours, an 8-car train that could accommodate as many as 800 people, seated and standing. Each car required only two 50-horsepower motors instead of the two 125-horsepower motors with which motor cars had been outfitted. No matter how many cars, only one motorman would be required. This was an opportunity both to enhance carrying capacity and reduce operating costs that no other elevated railway could ignore, and the MU system would also become standard with all subways, beginning with the IRT.

Boston and Philadelphia would build subways designed like New York's, and eventually, in the 1930s, ground was broken for subways even in Chicago, beneath State Street and Dearborn. Unlike the first London Underground, the Farrington–Paddington line, or most of the lines in New York, Boston, and Philadelphia, Chicago's were built by deep boring behind a "shield" that was forced ahead by hydraulic rams as iron rings were bolted together in its wake to form a continuous lining or tube. This technique was far more difficult and costly than cut and cover, but less disruptive of underground utilities, which by the 1930s had grown exceedingly complicated in any metropolis the size of Chicago.

PUBLIC OWNERSHIP AND FEDERAL FUNDING

The construction of new subways, along with the conversion of surface lines from trolleys to buses, afforded an opportunity to establish routes that better served shifting centers of population. Transit patronage in the United States surged during the early 1940s as gasoline was rationed and automobile manufacture ceased, reaching an all-time high in 1944 and 1945. After that, patronage declined at a breathtaking rate—well over 50 percent in just a decade before 1955, and more slowly but still steadily after that. As a private enterprise, mass transit had failed. Before World War II, only a handful of major transit systems had been turned over to public ownership, or, as the expression went, been "municipalized"—San Francisco's in 1909, Seattle's in 1911, Detroit's in 1922, the greater part of New York's in 1940, and Cleveland's in 1942. These five properties accounted for about one-fifth of the nation's transit ridership, and with the addition of Chicago in 1947 and Los Angeles, Oakland, and Sacramento in the 1950s municipal ownership still accounted for less than half. By 1960, however, the establishment of public transit authorities was generally regarded as the only way to forestall bond foreclosure everywhere else, and to salvage private investment—and, in many places, as the only way to assure the survival of mass transit at all.

In some smaller towns, the old companies simply went out of business with no replacement.

Municipal government had played a role in financing rapid transit in New York and Boston at the turn of the century, and surface transit was subsidized in Newark, New Orleans, Seattle, San Francisco, and Oakland, but the overall weight of public policy—at all levels, local, state, federal—had been thrown in favor of motorization even before World War I. Beginning with the Rural Roads program and culminating in the Interstate Highway Act of 1956—from which half of the funding went to urban freeways— public monies poured into highway construction whereas mass transit was held captive to the whims of the marketplace, required to fund itself and to pay taxes. Finally, it became clear that a crisis was at hand.

In 1961, the Institute of Public Administration in New York prepared a report on "Urban Transportation and Public Policy" at the behest of the U.S. Department of Commerce and the Housing and Home Finance Agency. Then, a year later, President John F. Kennedy called for the establishment of a program of federal capital assistance for mass transit:

> To conserve and enhance values in existing urban areas is essential. But as least as important are steps to promote economic efficiency and livability in areas of future development. Our national welfare therefore requires the provision of good urban transportation, and the properly balanced use of private vehicles and modern mass transport to help shape as well as serve urban growth.

And 2 years after that, the Urban Mass Transit Administration (UMTA—today's Federal Transit Administration) began providing capital grants for metropolitan areas that had developed comprehensive transit plans, and what came to be called the "transit aid lobby" (as a counterpart and foe of the "highway lobby") coalesced first around the problem of deteriorating railroad commuter service. It gathered force with the rise of environmentalism and organized resistance to new urban freeways. As it turned out, there would be no elevated superhighway along San Francisco's Embarcadero to link the Bay Bridge and the Golden Gate Bridge, and no 8-lane highway cutting a swath through the very heart of Washington, D.C., both of which were well along in the planning. Instead, there would be the Bay Area Rapid Transit (BART) and "the Great Society subway," the Washington Metro, which would eventually become the nation's second-largest and second-busiest rapid transit system and would remain "the most ambitious effort ever made to offer Americans an alternative to the automobile" (Schrag 2006, 1).

Metro's funding came from the federal government, and after 1973 some of the monies for funding rapid transit were even diverted from the theretofore untouchable Highway Trust Fund, which accrued from gasoline taxes. Typically, federal funds would cover 75 percent of construction costs, but of course the allocation of funds was always subject to the precepts and prejudices of those who had their hands on the levers of political power. There were those who applauded the government taking a direct role in promoting "an alternative to the automobile" and others who decried it. In 1986, Senator William Proxmire of Wisconsin bestowed his storied "Golden Fleece" award for wasteful government spending on UMTA for "playing Santa Claus to the nation's cities." There certainly were some spectacular flops, such as Detroit's automated People Mover, which never amounted to more than an UMTA-sponsored amusement ride, and never covered more than 20 percent of its operating costs. On the other hand, it was hard to argue against the transformation wrought by the Washington Metro.

Actually, there were many precedents for federal engagement with urban mass transit that long predated UMTA, a few of which had a positive effect, others of which implicitly tipped the scales further in favor of motorization. On the one hand, there was the $23 million in Public Works Administration (PWA) monies that helped pay for Chicago's new subways in the 1930s and 1940s, and there were also PWA funds in New York's Sixth Avenue subway. On the other hand, there was the damage done by New Deal antitrust legislation. During his first presidential campaign Franklin Roosevelt had called public utility holding companies "a kind of empire within the nation," and vowed to limit their influence. A 1935 federal law largely prohibited electric power companies from operating street railways as subsidiaries, which a good many of them had been doing. This cast many urban transit operations adrift from their relatively prosperous patrons and hastened the pace of abandonment. In the same vein, the purchase of PCCs was thwarted by the reluctance of the Reconstruction Finance Corporation (RFC) to underwrite loans for any mode of transit without rubber tires, because road-building was one of the primary New Deal employment projects nationwide. Only Kansas City Public Service ever managed to secure RFC funding for new streetcars, and in 1940 Seattle's municipal transit system was granted a large RFC loan only after it agreed to replace every one of its old streetcars, not with new PCCs but rather with trackless trolleys and gasoline buses.

The National Industrial Recovery Act nearly blocked the purchase of PCC cars altogether by stipulating "fair competition codes" that were exempt from antitrust laws and required individual manufacturing companies to adhere to "industry bids." Prices were purposefully fixed high, with the laudable aim of reviving industry and putting more people to work. But when the Brooklyn & Queens Transit (B&QT) first solicited a bid for PCCs from the St. Louis Car Company in 1935, the NRA codes were operative and the price was pegged at $24,000 each, at least $9,000 more than transit firms involved in the development of PCCs had expected to pay. B&QT's general manager tore up the bid and threw it out the window. Had not the codes eventually been declared unconstitutional, it seems likely that Brooklyn, like Manhattan, would never have had modern trolleys.

One can easily multiply examples of technological choice being contingent on the political process, specifically federal involvement in financial matters. During World War II, the War Production Board placed heavy restrictions on the acquisition of new PCCs, even though transit companies were prospering and could arrange their own financing. One exemption went to Washington, D.C., as streetcar patronage in the capital city more than doubled to 336 million annually, and Capital Transit was authorized to take delivery on PCCs all through the war. Another exemption went to Johnstown Traction in 1945, authorization to order seventeen PCCs with all the extras. By this time, nearly every town of Johnstown's size (60,000) was looking to disinvest in its streetcar infrastructure, and a decade later unemployment was rampant in Johnstown. To try and put a little cash in the coffer, the general manager put the 13-year-old PCCs up for sale while keeping older cars to maintain minimum levels of service. Even though the PCCs were far from fully depreciated, no buyer ever appeared and they were eventually scrapped, except for their electrical controls, which were shipped to Belgium for reuse with new Brussels trams. As for Johnstown Traction, it seemed clear that it would have fared better without the drain on its finances resulting from that RFC exemption.

A TURN OF FORTUNES

In the closing decades of the twentieth century, federal funding enabled construction of heavy-rail rapid transit in several cities that had never had such systems at all, not only BART and Washington Metro, but lines in Baltimore and Miami, even in Atlanta and Los Angeles, both of which had been totally reshaped after World War II by America's automotive culture. A majority of the world's great cities, including at least a dozen in North America, now have at least one subway line and many have networks. Indeed, any metropolis is *expected* to have a metro even though the costs have become astronomical. Depending on how one counts, the 103-mile Washington system cost between 10 and 20 billion dollars. On a cost-per-mile basis that was actually cheap compared to Los Angeles, where—after decades of aborted planning—construction finally started in 1986 on a subway that would connect Union Station with the San Fernando Valley. It ended up costing 5.5 billion dollars even though it was only 17.4 miles long.

There were ongoing protests in Los Angeles and elsewhere that the tradeoffs were too steep, that new subways were too costly for the actual benefits provided, especially since the priority given to high-profile high-tech projects was starving monies that would better be spent upgrading ordinary bus lines with new equipment, or even fixing broken sidewalks. And yet the presence of a subway became so intertwined with urban self-image and notions of civic pride that such projects seemed to have a momentum of their own. Nevertheless, it now appears that most of the new North American subways are likely to remain only a fragmentary reflection of the great systems built a century ago in New York, London, and Paris, when part of the expense could reasonably be borne by private capital. There has been talk about the line in Los Angeles being "the last subway." Whether or not that turns out to be true, there is every likelihood that new rail-transit initiatives will be a less expensive variant that is not wholly isolated from other traffic and therefore does not use an electrified third rail. On the surface, light rail often looks very much like heavy rail, as is suggested by the first illustration in this book. Yet, light rail vehicles (LRVs) are distinctly reminiscent of streetcars—most notably because the current is transmitted through overhead wires so that passengers can board safely from the street (Figure 5.2).

LIMITED TRAMLINES

The form of urban transport that became known as light rail had a prehistory that predated the availability of funding from UMTA. Most of the American trolley lines that survived beyond the 1960s had rights-of-way that were at least partially isolated from automotive traffic. In Europe these were called "limited tramlines." A notable example in the United States was a 15-mile line completed in 1920 as the Cleveland Interurban Railroad by the brothers Oris P. and Mantis J. Van Sweringen to provide transportation to and from a new a residential subdivision, Shaker Heights, Ohio. Later it was called Shaker Heights Rapid Transit, which was a misnomer—almost the entire line was on the surface—although much of it did run in the broad center median of two boulevards, Shaker and Van Aken, and it managed to survive with PCCs even as all of Cleveland's other trolley lines were being converted to diesel buses.

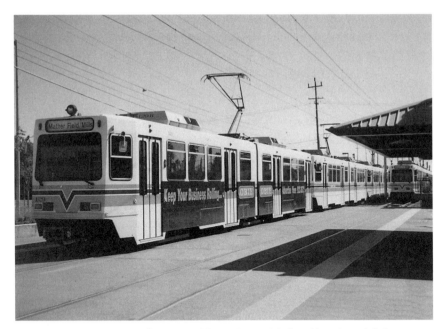

Figure 5.2: In Sacramento, the Regional Transit District's light rail lines largely follow former railroad alignments, as do light rail lines in many other American cities. But in some places, including the vicinity of the state capitol building, Sacramento's LRVs literally become "streetcars." There are several technological distinctions between the newest LRVs and traditional streetcars, however: compact solid-state electrical components take up less space underneath—so cars can therefore have lower floors—and on-board inverters change the direct current (DC) from the overhead wires to 3-phase alternating current (AC). This photo shows another distinction; the cars are articulated, with a truck on each end and another supporting the jointed section in the middle. Passengers validate a ticket inside the car after boarding, a technique borrowed from European practice. (Author's photo)

There was much the same situation in San Francisco and Pittsburgh, both with lengthy streetcar tunnels, Twin Peaks and South Hills; New Orleans with its neutral ground and Mexico City with its *tierra reservada*; and Philadelphia, Newark, and Boston with downtown trolley subways. The first extension of a North American trolley operation in nearly 40 years took place in 1959 when the Massachusetts Bay Transportation Authority took over an abandoned railroad right-of-way, installed 9 miles of new track from Brookline to Newton, and routed PCCs in and out of Boston's streetcar subway. Service was much quicker and more convenient than any over-the-road alternative to and from Boston's burgeoning western suburbs, and the line attracted substantial patronage from the start.

Boston's new line, called the Riverside Line, provided persuasive evidence for the potential of limited tramline service to attract patronage if it could beat the time it took commuters to drive their own cars. Even more persuasive was the evidence provided by a project initiated by the Chicago Transit Authority's director of planning (and later executive director), George Krambles (1915–1999). In the early 1960s, Krambles set in motion a plan to rehabilitate the former Chicago, North Shore & Milwaukee line connecting with the village of Skokie on Chicago's northern city limits—again, an area of rapid suburban expansion. This time, there was a token amount of federal aid through the Department of Housing and Urban Development. The "Skokie Swift" was billed as a HUD "demonstration project" (Figure 5.3). What it demonstrated, in Krambles's words, were the benefits to be realized from . . .

> encouraging the investment of taxpayer revenues in judiciously planned rail transit, particularly in relatively low-cost, no-frills service using rail corridors whose original purposes have faded over the years; the kind of service that became identified a few years later as light rail transit.

Krambles in now called "the father of light rail." It took a few years, but in one or another variation the Skokie Swift provided a model for new electrified transit lines in eight other North American cities during the 1970s and 1980s, many more than that since 1990, and now even on new lines in Great Britain and France.

FULL CIRCLE

Except for the St. Charles Avenue line in New Orleans, which continued to operate ancient heavyweight trolleys dating back to the 1920s, every streetcar line that survived into the 1970s used PCCs. By then, even the newest of them was near the full extent of its planned service life, and UMTA was investigating possibilities for subsidizing the development of a Standard Light Rail Vehicle (SLRV) with the aim of repeating the success of the PCC. Eventually, a contract for developing a SLRV went to an aerospace firm that was looking to broaden its product line in the wake of defense-spending cutbacks, Boeing-Vertol, which had been in the business of building helicopters for the armed forces and was used to operating without much in the way of cost constraints. Boeing-Vertol's engineers came up with a 71-foot long articulated with six axles, two 230-horsepower

Figure 5.3: The Skokie Swift, which began operation in 1964, was light-rail transit before the name was invented. The cars were built using parts salvaged from PCCs and the line provided a connection between the Chicago Transit Authority's heavy-rail system in northern Cook County and the village of Skokie, near O'Hare Airport. (Chicago Transit Authority publicity photo, author's collection)

motors, solid-state DC control, acceleration even better than a PCC's, and a top speed of 55 miles per hour—a high performance vehicle in every way. But in operation they were a headache from the start; expertise in aerospace engineering did not transfer readily to mass transit. They were complicated and expensive to maintain, there were electrical problems, problems with the brakes, the doors, and the point of articulation. After the initial 250-car production run was divided between Boston and San Francisco, Boeing-Vertol went out of business.

More successful than the UMTA project was a vehicle developed for the Toronto Transit Commission, the Canadian Light Rail Vehicle, which also found a market in the United States, on a new light-rail line in San Jose, California. As work was under way on BART and other heavy-rail systems, it was beginning to appear in many cities that such systems were too daunting in cost/benefit terms; Santa Clara County, where San Jose is located, elected not to get involved with BART. On the other hand, limited tramlines on the order of the Skokie Swift were showing the feasibility of a transit option that was less time-consuming, less disruptive, and in particular less voracious of capital investment and more readily funded at least in part through state and local sources. Such lines might have high-level platforms in order to expedite dwell time, but what they all shared in common was at least some operation that was *in the street*, with electric current coming from overhead wires.

The first all-new North American light-rail line was opened in the oil-rich city of Edmonton, Alberta, in 1978, and another in Calgary in 1981. But the light rail revival in the United States that same year had a special irony, for it took place in San Diego, which had been the very first city with PCCs to throw in the towel and convert its transit entirely to buses, back in 1949. For their new lines, Edmonton, Calgary, and San Diego all bought what became the closest thing to an LRV that was standardized to the same degree as a PCC car had been—accounting for almost half the 1,600 LRVs delivered to North American transit systems during the 1980s and 1990s. It was a double-truck design that had initially been developed for service on tramlines in Frankfurt, Germany. Larger and more powerful articulated versions later went to new light-rail operations in Sacramento, Denver, Salt Lake City, and St. Louis, and also to Pittsburgh in order to replace PCCs on the few surviving trolley lines operating through the South Hills Tunnel. The manufacturer? It was almost as if the history of the streetcar had come full circle: Final assembly of these cars was completed in California, but the manufacturer was headquartered in the German city of Düsseldorf. And the firm traced its lineage back to Ernst Werner von Siemens, the inventor who in 1879 had first figured out how to transmit electric power from a stationary dynamo to a moving conveyance, and thereby solved the major technological problem beyond which lay the urban transit revolution wrought by the streetcar. The firm was called Siemens-Düwag (Düsseldorfer Waggonfabrik).

6

Conclusion

Along with Siemens-Düwag coming to the forefront in the manufacture of light rail vehicles, the revival of fixed-guideway urban transit in North America presented quite an array of ironic full-circle scenarios. Another is worth mention in concluding. After PCCs disappeared from San Diego in 1949, the only trolleys left anywhere in Southern California were those that ended up under the control of the Los Angeles Metropolitan Transit Authority, and in the early 1960s LAMTA substituted buses on the last of the lines once operated by the Pacific Electric and the Los Angeles Railway. After that, it was all but certain that trolleys were gone forever. Even though a county agency was mandated to implement new rail-borne transit, the odds against this seemed overwhelming, especially after the repeated failure of ballot initiatives to secure extraordinary funding. For those charged with running the transit system in the face of declining patronage and escalating costs, rubber tires on public roads—buses with diesel engines—were the only option they could afford. But we have already seen how the fortunes of technological systems are often a matter of winds in ambient political realms, and in the 1980s unfavorable winds turned fair when enthusiasts for a rail revival managed to draw on positive (if erroneous) perceptions of the old PE Red Cars, not so much for what they were in and of themselves, but as a symbol of a safe and snug past. The result was that voters affirmed a half-cent general sales tax to help finance new rail-borne urban transit, the funding needed to insure what one of the newspapers called "the most dramatic turnaround since Los Angeles' enduring commitment to freeways." Thirty years after the last Red Car made its last run between Los Angeles and Long Beach, a brand new *Blue* Line was completed between those same two cities, following almost exactly the same route.

Not all of the new heavy rail and light rail lines—those in Miami and Buffalo, for example—have measured up to patronage forecasts that were overly enthusiastic, or, some would say, deliberately exaggerated. One of the four Los Angeles lines, the Green Line that runs in the median of the Century Freeway—the very last segment of the Interstate Highway System—has also been a disappointment. Partly this is because it was intended to serve thousands of workers in an aerospace industry that all but collapsed in the 1990s (shades of Johnstown, Pennsylvania, in the 1950s with its new PCCs and an imploding steel industry). But the Blue Line, retracing the old Red Car route to Long Beach, has become the best patronized light-rail line in the country, and similar lines in San Diego, Dallas, Denver, Sacramento, St. Louis, Salt Lake City, and especially Portland, Oregon, are also success stories. At the turn of the twenty-first century, there were only a few metropolitan areas in the United States that had no rail-borne transit, most notable being the Motor City, Detroit (see Table 6.1). Surely Detroit's local transit system is the worst of any big city's. But with projected costs for light rail lines routinely reaching toward a billion dollars it remains an open question whether the list of cities with LRVs will get much longer, at least in the near future.

It should go without saying, however, that the future is by definition uncertain: When "hopelessly old-fashioned" trolley systems were sputtering out of existence 50 years ago, who would have believed that a reincarnation of trolleys in the form of LRVs would ever be sparked by fond memories (or, rather, fond images) of the way urban mass transit was in the past? Or that LRVs, providing transit on rails with power from overhead wires, would be seen as an emblem of urban modernity?

For much of the twentieth century, the main problem that urban transit was designed to alleviate was *congestion*—too many vehicles and too little space. "It would undoubtedly be the ideal standard of travel," wrote Philip Harrington, head of the Chicago transit system in 1940, "if everyone were able and could afford to execute all his own movements in his own car, in complete safety, free from all traffic annoyance and delay." Many years later, and partly because of "traffic annoyance and delay," mass transit would enjoy a resurgence. And yet personal mobility remained an ideal for which most Americans—and, increasingly, people in other parts of the world—would accept many tradeoffs, all other things being equal. But there is the

Table 6.1
North American light rail lines, 1978–2006

City	Operator	Service began
Edmonton	Edmonton Transit System	1978
Calgary	Calgary Transit	1981
San Diego	Metropolitan Transit System	1981
Buffalo	Niagara Frontier Transp. Auth.	1984
Portland	Tri-County Metropolitan Transp.	1986
Mexico City	Servicio de Transportes Electricos	1986
Sacramento	Sacramento Regional Transit	1987
San Jose	Valley Transp. Auth.	1987
Guadalajara	Sistema de Tren Electrico Urbano	1989
Monterrey	Metrorrey	1991
Los Angeles	Metropolitan Transp. Auth.	1991
Baltimore	Mass Transit Administration	1992
St. Louis	Bi-State Transit System	1993
Denver	Rapid Transit District	1994
Dallas	Dallas Area Rapid Transit	1996
Salt Lake City	Utah Transit Authority	1997
Jersey City	New Jersey Transit	2001
Minneapolis	Minneapolis Transit	2004
Houston	METRO	2004
Streetcar lines with light rail vehicles		
Pittsburgh	Port Transit Auth.	1891
Toronto	Toronto Transit Commission	1892
San Francisco	Municipal Railway	1897
Boston	Massachusetts Bay Transp. Auth.	1897
Philadelphia	Southeastern Penn. Transp. Auth.	1905
Cleveland	Greater Cleveland Regional Transit	1920
Newark	New Jersey Transit	1935

catch: In our own time it seems increasingly unlikely that all other things will remain equal, and especially not an abundant and cheap supply of the liquid hydrocarbon that fuels automobility.

When electric transit regained a long-lost momentum in the 1970s, this was in the context of a new concern about the environment and a new apprehension that personal mobility was threatened by recurrent "gas crises." Now, 30 years later, environmentalism has taken on heightened urgency in the face of overwhelming evidence for global warming, and personal mobility appears to be threatened by a *permanent* gas crisis, as international rivalry escalates for an earthbound supply of petroleum that surely is finite. This may again impart momentum to the construction of new electric transit lines, though almost certainly not of the magnitude of projects like BART or Washington Metro, or the other systems elsewhere that have been designed not just as facilities for transportation but as monuments: So much money was blatantly wasted on the subway line in Los Angeles that voters actually passed an initiative expressly forbidding any more subways. Which is not to say that money has not been very well spent on more modest projects, the nearby San Diego Trolley being a prime example, not to mention the Blue Line to Long Beach.

So, will there be more light rail? More American cities with electric railways? In 2000, there were more cities with electric rail than there had been 50 years before, and yet reservations will persist about the cost of capital-intensive infrastructure, even the need. So, what about ... *buses*? Not just ordinary buses and bus lines, but rather express buses with real-time information systems, "high-tech passenger amenities," transponders that favor them with green lights at intersections. As far back as the early 1970s, cities had begun to establish "bus malls" and bus *lanes* on downtown streets and on urban freeways (Figure 6.1). At a higher level of technological sophistication, much fanfare has attended the institution of entire *lines* that are free of "mixed traffic" from end to end, except at intersections, lines capable of providing what is called "bus rapid transit" (BRT) in the United States, or, less pretentiously in Europe, "quality bus service." Boston has its Silver BRT Line, the San Fernando Valley its Orange BRT Line, and it is a pretty safe bet that there will be more such lines elsewhere.

Nevertheless, as the cost approaches the cost of light rail—$324 million in the case of the 14-mile Orange Line linking North Hollywood and Canoga Park—it is likewise certain that nothing of the sort is ever going to be feasible for the vast majority of transit operations, now 500 of them in U.S. cities and towns alone, and another 1,200 in rural areas. It may be feasible in Los Angeles, where the transportation authority carries 375 million bus passengers annually, or Chicago with 292 million, but definitely not feasible for MATS in Montgomery, Alabama, with its eighteen diesel buses and its twelve paratransit vans—or, to get even closer to the grassroots, for MUST (Maryland Upper Shore Transit), with its potpourri of handicap-accessible paratransit vans. The service provided by MATS and MUST is no less essential to the regions they serve, but there will be no 6-axle diesel-electric hybrids from the Czech Republic, no Volvo du Brasil triple-articulateds fueled by compressed natural gas (to mention only a couple of the ultra-high-tech buses that one finds advertised in *Jane's*).

Rather, for operations like MATS and MUST—and for most others, even operations that are much larger—there will be plain old diesel buses quite similar to those that once saved the transit industry from utter collapse.

Figure 6.1: In the 1970s, the Chicago Transit Authority converted State Street into a "bus transit mall," an idea that seemed to have promise at the time. The problem was that buses still had to fight their way out into regular "mixed" traffic as they exited their mall, and gains in patronage were negligible. The outcome with newer "busways" in several cities has been more favorable, although Los Angeles recently terminated a 32-year "experiment" with what is called *contraflow*—setting aside a lane for northbound bus traffic on otherwise southbound Spring Street. (Chicago Transit Authority publicity photo, author's collection)

One truand secodthese will not be the undermaintained "third-world buses for third-world people," that have sometimes been prevalent in cities with glittery showpiece subways that actually handle only a fraction of the total transit load. One also trusts that they will not pollute the air as dreadfully as first- and second-generation diesels did for so many years (and the great majority of the 500,000 school buses in the United States still do). But if transit ridership increases significantly in the next few years—as is more than likely, assuming that there will never again be cheap gasoline—the workhorse is almost certain to be much the same sort of vehicle that has dominated urban transit for even longer than the streetcar it superseded.

We can lament the electric railways that were abandoned when they should have been saved, not an insignificant number of them. We can lament the deficient commitment for so many years to serious investment in new electric railways—or, rather, contrast what happened in most of the "advanced" countries of the world to what happened in others, many of them in the Soviet Union, during the two decades after World War II. We can lament that the operation of diesel buses is contingent on the same finite supply of petroleum as automobiles are, while at the same time pointing out the obvious disparity in how much fuel it takes to impel a busload of people compared to a procession of automobiles—everyone executing "all his own movements," as Mr. Harrington put it. And if it turns out that heavy rail, LRVs, ETBs, busways, hybrids, LNG, CNG, and hydrogen fuel cells are mostly beyond reach in all but a handful of the largest cities with the most political muscle, and that the plain-vanilla diesel bus really is the future of urban transit nearly everywhere, this will perhaps be the final and best example of history repeating itself. "The past is never dead," wrote William Faulkner. "It is not even past."

Afterword: But What About Judge Doom?

The story is essentially this. Before improved roads, before mass production, before the jitney and the inexpensive used car, there was a system of transit that flourished in every U.S. city and every town of any size, and indeed throughout much of the Western world. At the heart of this system was a singular conveyance, the electric streetcar, more than 70,000 of them in the United States alone, plying 40,000 miles of track, not even counting the interurban electrics that fanned out through the countryside, linking towns to cities that provided "hubs," not unlike the later practice of the airlines. Whether urban or interurban, electric railways enjoyed one particular advantage: A great many individual public conveyances, dozens, hundreds, could derive their power from one central powerhouse. This was in complete distinction to the railroads, where every train had to have its own locomotive, or indeed to motor vehicles, where every conveyance was likewise self-powered. From before the turn of the century until the 1940s, trolleys were what moved the masses.

Although the national government had appropriated funds for turnpikes in the early days of the republic, no such financing had been made available since the 1840s. In 1907 only 7 percent of the roads in the United States had any surfacing at all, and that was usually just gravel. Then in 1916 Congress voted to allocate $75 million over a 5-year period for the improvement of rural post roads. This was followed in 1921 by legislation that provided matching grants to states, in order to promote an interconnected national system of highways. Downtown streets were being improved in large measure at just the same time that many more families could budget for an automobile. And also at just the same time, street railways were losing profitability in the face of inflation, while it was proving politically difficult either to cut labor costs or increase fares. Profits revived briefly in middle 1920s, then again began to wane as the market was flooded with used cars. Reduced patronage as a result of the Great Depression hit the smaller transit systems hardest, but even big-city systems were sent into bankruptcy. People had moved to the suburbs because the street railways made it possible to get back and forth—made it easy, actually, despite continual complaints about inadequate service. But after a while, more and more people found it preferable to get back and forth on their own, in their own autos. It might even be said that urban transit became the victim of its own success.

There were far fewer U.S. cities and towns with street railways in 1940 than there had been in 1890, and two hundred of them had *only* rubber-tired transit vehicles. Although there would be a major resurgence in patronage during World War II, for small systems good times were gone forever. In the larger cities PCC trolleys, along with trackless trolleys, fought a rear-guard action for 10, 15, sometimes 20 years after the war. By the mid-1960s, however, the situation was the same in all but a handful of cities: Commuters who used public transit rode buses, and those who did not—the overwhelming majority in most places—drove automobiles. The trolleys that had so completely dominated the urban panorama were almost gone, a small number of them sold overseas, most reduced to scrap (Figure A.1). In fewer cities than one could count on one's fingers there remained remnants of the vast networks that had once crisscrossed each other on all of the downtown streets and served all of the old suburbs. The urban panorama had been profoundly transformed.

Transformed is the operative word. Urban transit had not evolved in any preordained way, nor is the term "evolution" even appropriate with reference to technological change. Nor is the term "progress," which always begs the question, "progress for whom?" More often than not, technological change is the outcome of decisions made by people who have the political power to determine where financial resources will be concentrated or where they will be withheld or withdrawn. One can sometimes read something sinister into this process, as when turn-of-the-century trolley barons conspired with corrupt municipal officials to gain special privilege in return for favors bestowed. But whether sinister or not, technological change remained contingent on human agency that was often exerted through the political process.

Figure A.1: As trolley systems skidded toward oblivion, one of the most memorable images was this Terminal Island scrap-yard, another photo of which was published in *Life* magazine in 1956. Stacked up like cordwood are streetcars that had plied Hollywood Blvd. beginning in the 1920s, while in the foreground are the remains of a Peter Witt car ordered by the Los Angeles Railway to herald a modernization program that was cut short by the Great Depression. (Author's photo)

Many choices were involved in the triumph "motorization," from minimizing gasoline taxes to maximizing investment in roads and highways. Beyond such choices, explanations for motorization ranged far and wide. They could be put in the most judgmental of terms: Some said that street railways were "water-logged in the mire of financial filth," while buses were "young and honest" (quoted in Schrag 2000, 52). For sure, once buses reached a certain stage of technological development, the people who ran transit lines began to regard their economics as preferable to the economics of trolleys, in the United States, anyway. But in the United States were the scales somehow tipped? A history of one of our major transit companies is titled *Who Made All Our Streetcars Go?* (Farrell 1973). The question has a studied ambiguity. Who made them *go* can be taken to mean who made them *run* and to that question one would answer John Stephenson, Andrew Hallidie, Frank Sprague, and many others. But it can also be taken to mean who made them *go away*? Here, one might mention Henry Ford and Charles Kettering, among others. This book has also invoked many forces that are not so readily personalized, such as monetary inflation and labor policies, few of which have much dramatic appeal. But there is another way to answer the question "Who made them go away?" This answer comports with a popular fascination with the "back story," it is freighted with a rich array of symbols and metaphors, and it conforms to a perception that carries the immense power of nostalgia.

What really happened when motor vehicles superseded trolleys and the technology of urban transit was utterly transformed? Shifts of this magnitude rarely admit of simple explanations, and yet simplicity has an abiding appeal. As people pondered this transformation, and as time passed, many of them found it hard to believe that it was solely the result of changing economics or better roads or improved buses or anything of the sort. They tended to forget that traction companies had once been regarded as predatory monopolies, and that, later, an overcapitalized transit industry had been swept by bankruptcies. What they saw instead were evildoers who collaborated to strip urban America of a superior technology for no good reason except to enhance their wealth and power. They saw a tragedy, and at its root what they first saw was a swindle, and then they saw a conspiracy.

Word of a swindle was first broadcast widely in 1946, in a 24-page manifesto that showed up in the mail of municipal officials and transit engineers all around the country. It began: "This is an *urgent warning* to each and every one of you, that there is a careful, deliberately planned campaign to swindle you out of one of your most important and valuable public utilities, your Electric Railway system." This manifesto was the work of Edwin J. Quinby, who had formerly worked for an electric railway operating from Paterson, New Jersey, to Suffern, New York, and in 1934 founded an organization called the Electric Railroaders' Association. The ERA was dedicated to preserving and publishing historical information on electric railways (its efforts have been essential to the writing of this book), but it also had an explicit political agenda, to stand firm on their behalf wherever there was the threat of abandonment in favor of buses.

In the mid-1930s, it was just becoming clear that the switch from trolleys to motorbuses was a strategy that was infecting the management of systems in larger and larger cities. By 1946 there were nearly twice as many buses in urban transit as trolleys, and some of the very largest cities seemed to be headed for complete motorization, or, in the coinage of the ERA, for "bustitution." To Quinby and his ERA cohort, such a turn of events seemed terribly irrational. But they knew it must have been rational from *some* standpoint—rational and also criminal, hence perceptions of a "swindle." They understood that transit systems in quite a number of cities—not just small cities like Montgomery but also in Philadelphia, Baltimore, Oakland, and Los Angeles—were under the control of National City Lines, which was an instrument of General Motors and provided transit lines with capital for what it termed "modernization." But they did not fully connect the dots that would turn a swindle into a conspiracy.

That fell to a young staff lawyer who testified at length to the U.S. Senate's Subcommittee on Antitrust and Monopoly in the winter of 1974. His name was Bradford Snell (b. 1946). The committee was chaired by Michigan Senator Philip Hart, who was explicitly interested in breaking up General Motors, the most powerful corporation in the world, its net value greater than all but a handful of sovereign nations. Senator Hart wanted to force divestment of its largest automobile division, Chevrolet, and also Electro-Motive, the division that had produced most of the diesel railroad locomotives that replaced steam throughout the United States after World War II. Snell had apparently investigated GM's monopolistic practices in exhaustive detail, and he had prepared an imposing report with hundreds of footnotes, a report with an air of unfailing authority. The report addressed the questions that Senator Hart was concerned with, but it also included much information on National City Lines, and the nature of its control by GM, along with Standard Oil of California and Firestone Tire & Rubber.

According to Snell, as soon a federal antitrust legislation cast transit systems loose from electric power companies in 1935, NCL began getting control of such systems with the immediate aim of scrapping the streetcars and financing the substitution of buses. But that was not the *ultimate* aim: Because there was no doubt, he said, that patrons would find buses vastly inferior to streetcars, they would soon elect to start driving their own automobiles, more and more and more of them.

Even though Senator Hart failed to realize his ultimate aim—legislation compelling GM to divest key corporate components—Snell's allegation that electric railway systems had been deliberately destroyed by marauders from the automotive world took hold in the popular imagination. The tale was told and retold in highbrow periodicals like *Harpers*; in monographs by eminent historians; on *Sixty Minutes*; in a PBS documentary called *Taken for a Ride*, in which the narrator tells how "smooth, clean, and comfortable streetcars ruled America's cities" until General Motors stepped in and left us with a "dystopian nightmare." And then in 1988 Hollywood released a remarkable movie, *Who Framed Roger Rabbit*, which dramatized the conspiracy in an unforgettable scenario. During a confrontation just after World War II in a part of Los Angeles called Toon Town, the murderous Judge Doom revealed to private detective Eddie Valiant his "epic plan" for transforming metropolitan Los Angeles:

"They are calling it a freeway," says Doom.

"Freeway? What the hell's a freeway?" asks Valiant.

Doom's voice rises. "Eight lanes of shimmering cement running from here to Pasadena. Traffic jams will be a thing of the past."

When Eddie protests that "nobody is going to drive this lousy freeway when they can take the Red Car for a nickel," Doom assures him, "Oh, they'll drive. They'll have to." Eddie's jaw drops as Doom explains that he has bought the streetcar system *so he could dismantle it.*

For "Judge Doom" read General Motors. For "Red Car" read Pacific Electric, Southern California's once proud and still famous network of electric railways. According to Bradford Snell, dismantling the Pacific Electric was the worst of the many crimes committed by General Motors through the instrument of National City Lines.

It so happens that presently, early in the twenty-first century, General Motors looks like a giant enfeebled. But for many Americans 30 years ago, you could not have invented a more compelling corporate villain. GM's CEO, Charles Wilson, was thought to have said that "what's good for General Motors is good for the country," when it actually appeared that exactly the opposite was true more often: The exhaust from GM diesels certainly was not good for anybody. As the Quinby/Snell/Judge Doom explanation of motorization was elaborated and gained credence, it occasioned little dissent. General Motors issued a rejoinder, of course, but the one really credible voice was that of George W. Hilton, an economist whose credentials included chairmanship of President Lyndon Johnson's task force on Transportation Policy, a shelf of scholarly monographs in transportation economics, and another shelf of histories, including the definitive histories of interurban electric railways and cable railways that have provided a good part of this book's source material. Hilton regarded the demise of the Red Cars, like the demise of streetcars throughout the United States, as merely a shift in public preferences in the context of new economic circumstances. More and more people preferred driving their own cars to riding public transit. Ultimately, the PE's management "quite properly let the equipment run down until it became fully depreciated," its relics symbolizing an obsolete technology passing naturally into the hereafter.

But Hilton's was a voice in the wilderness. The tale of conspiracy became, and to a large extent remains, part of our conventional understanding of the past: A superior form of mass transit fell victim to corporate miscreants, who substituted something altogether inferior in the expectation that mass transit would fail altogether and everyone would be forced to drive automobiles. Any reader who doubts the hold that this scenario has on popular consciousness is invited to try an experiment the next time he or she is part of an informal social gathering: Simply announce that traffic congestion is becoming intolerable, and then pose a question: Would not the situation be ever-so-much better if only General Motors had not destroyed our trolley systems? It is virtually certain that heads will nod in affirmation.

I trust that this book has presented an explanation of the trolley's decline and fall that is more complex if less dramatic than the purported machinations of conspirators like Judge Doom. Which is not to argue that the triumph of motorization was simply the result of free-market capitalism in action. Were the nabobs of the automotive world pleased by what happened? Of course they were. Did they hurry it along? They did what they could, not least by throwing their considerable political heft behind a highway lobby that realized an ultimate dream when President Dwight Eisenhower approved the largest public works project in history, the Interstate Highway Act of 1956. This paved the way for colossal expenditures on urban freeways just as automobile manufacture was becoming "one of the industrialized world's great growth industries" (Volti 2004, 89).

But the fate of electric railways is more contingent, more fraught with irony, and ultimately more interesting than the stark tale of corporate wrongdoing that has such a strong hold on the popular mind. Make no mistake: Buses *were* inferior in one significant regard. There is no doubt that the substitution of diesel power entailed immense tradeoffs in terms of degrading the quality of the environment. Yet make no mistake either about the reality that the transit industry was forced to confront: A former transit executive writes that

> Industry managers faced a near-impossible task, balancing the interests of investors who expected profits, employees who demanded better wages and working conditions, riders who wanted low fares and frequent service, and fickle politicians turned regulators who determined the fares streetcar companies could charge, the services they had to provide, and their return on investment—assuming, of course, there was any return at all (Diers 2006, 56).

Historians sometime suggest that the mass transit industry "collapsed" when buses were substituted for streetcars. Transit certainly endured a dire crisis, but just as certainly it is still a major presence in America's urban panorama: In 2005 the transit authorities in New York and Los Angeles each registered more than 1.5 billion passenger miles on their bus lines alone and even in cities as disparate as Honolulu and Minneapolis the figure is near 300 million.

As for the conspiracy theory, it is easy enough to find holes in the evidence. First, while the alleged conspirators did acquire local transit systems all across the country, and they did substitute buses, they never acquired the Pacific Electric's interurban system; the PE was *not their property to dismantle.* (Why Bradford Snell said it was is anybody's guess.) The alleged conspirators did purchase the Los Angeles Railway from Henry Huntington's estate, and they did substitute motorbuses on several lines. And yet they also carried out a program to equip the most heavily traveled lines with new PCCs and trackless trolleys in the late 1940s, and those lines were still served by trolleys and trackless trolleys when a public agency took over in 1958 (as public agencies eventually took over everywhere). Nationwide, the ultimate reach of the alleged conspirators extended to only about 10 percent of all transit systems—sixty-odd out of some six hundred—and yet virtually all the other 90 percent also got rid of trolleys (as happened with all the tramcar systems in the British Isles and France).

By focusing attention directly to "the freeway," Doom actually gets closer to the truth than did Snell. Rather than evildoers who set out "to reshape American ground transportation to serve corporate wants instead of social needs," a prime determinant was the repeated affirmation of the virtues of automobility. The publicist Bruce Barton extolled "the magic of gasoline" in the 1920s; 40 years later President Ronald Reagan called the private auto "the last great freedom." It was not Judge Doom (aka General Motors) who destroyed the profitability of mass transit and made our trolleys go away. Rather, it was *us,* all of us for whom private mobility became one of our most treasured prerogatives, even in the face of increasingly drastic tradeoffs—gridlock, for one, and 50-dollar fill-ups at the gas station. In analyzing technological change we need to not only take account of more-or-less rational factors such as cost/benefit analyses, but also human emotions—or the manipulation of emotions—such as pride or prejudice, sentimentality or enthusiasm. What makes the tale of conspiracy so interesting, if no less erroneous, is the way it plays to romantic images of a paradise lost. What makes the actual history of urban mass transit so interesting are all the tradeoffs that people have made and been willing to accept all along the way.

Glossary

All Service Vehicle (ASV). A bus capable of operating either with an internal-combustion engine, or with twin trolley poles extended, on electricity from a central generating station.

Alternating current (AC). A system of electrical distribution in which the current changes direction in regular cycles, the frequency of which is expressed in hertz (Hz).

Articulated. A transit vehicle with two hinged sections that are permanently connected and supported by a center truck, now used on most light rail lines.

Automated light rail transit. A system that relieves the operator of an LRV of performing any manual functions except in an emergency.

Auxiliary power unit (APU). A battery than enables electric trolley buses (trackless trolleys) to be driven "off wire" for short distances.

Average travel distance. A system used by transit planners to compute anticipated ridership on any given transit line.

Ballast. Crushed rock used for holding track in alignment.

Birney safety car. A standardized lightweight 4-wheel car designed for one-man operation, first manufactured in 1916.

Bobtail. A single-end horsecar with a step directly up to the door in the rear instead of a platform.

Bond. A cable brazed or welded from the end of one 42-foot section of rail to the next section in order to assure a continuous negative return for the electric circuit.

Brill. The J. G. Brill Co. of Philadelphia, manufacturer of streetcars, trolley buses, and other types of vehicles, 1868–1956.

Brilliner. A streamlined high-tech trolley debuted by the J. G. Brill Co. in 1938 in an attempt to compete with the PCC car.

Brush. The block of material, usually carbon, that conducts electric current from the fixed part of an electric motor (the stator) to the rotating part (the rotor).

Cable railway. A railway whose streetcars are impelled by a steel cable powered by a steam engine and moving continuously in a conduit beneath the tracks.

California car. A type of streetcar developed initially for the California Street Cable Railroad with a closed center section and open sections at either end.

Car. Synonymous with a streetcar or a trolley car.

Carbarn. A shed or building for the storage of horsecars, the name eventually designating a storage building for any kind of streetcar.

Catenary. Overhead power distribution system that embodies both messenger and contact wires. Contacted from cars below by means of a hinged device called a pantograph.

Conductor. (1) A crew member in a transit vehicle who signals orders to stop and proceed, collects fares, and keeps order. (2) Any material or device that conducts electric current with a high degree of efficiency.

Conduit. A trench between streetcar tracks in which there is either a moving cable or the electrified wire that ordinarily would be overhead.

Controller. The device that enables the operator of a trolley car to change speeds by changing resistance in the electrical circuit.

Convertible car. A type of streetcar in which side panels can be removed and stored during summertime operation and replaced in winter.

Crossover. Special trackwork that includes two switches facing in opposite directions and enables a streetcar or LRV to switch from one track to the adjacent track.

Cut and cover. A technique for constructing subways by digging from street level rather than "deep boring" underground.

Dead man pedal. A safety device which a streetcar operator must depress continuously or else the brakes will apply automatically.

Destination sign. A metal plaque or a backlit transparency indicating where a transit vehicle is bound.

Diesel engine. An internal combustion engine in which the fuel is ignited by compression alone, rather than by the introduction of an electric spark into the combustion chamber.

Double decker. A transit vehicle with two levels of seating, the top level reached via a staircase inside the car.

Double trucker. A streetcar with two 4-wheeled trucks, one under each end.

Direct current (DC). A system of electrical distribution in which the current flows in one direction only.

Dummy. A small steam locomotive used to impel streetcars, which has the boiler and running gear shielded by a wooden body, ostensibly to conceal its true nature.

Dwell time. The interval between the times a transit conveyance stops to load and discharge passengers until it is moving again.

Dynamic brakes. A system in which traction motors operate in reverse to provide braking.

Electric trolley bus (ETB). *See* trackless trolley.

Elevated railway (El or L). A rapid-transit infrastructure comprising long viaducts, usually but not always located over city streets.

Energize. The act of turning on the power to an electrical component or system.

Farebox. A receptacle into which a transit passenger deposits the fare directly rather than handing it to a conductor.

Fare register. A mechanical recording device on which a streetcar conductor is duty-bound to "ring up" each fare as it was collected.

Feeder. (1) The cable used to conduct electrical power to a trolley wire or catenary. (2) A short transit line, usually in an outlying area, that "feeds" passengers to a main line.

Fender. A device mounted at the front of a streetcar designed to protect pedestrians from being run over.

Fixed guideway. A trolley, light rail, heavy rail, monorail, or funicular used to guide conveyances in the absence of steering gear.

Flange. The protruding part along the inner rim of a wheel designed to keep it on the rail and guide it around turns.

Folding step. A step into a streetcar that flattens out when the doors are opened and retracts when the doors are closed.

Gauge. The distance between the rails measured from the inside face of one rail to the inside face of the other; "standard" gauge is 4 feet 8.5 inches, but narrower and wider gauges were not uncommon with streetcar lines.

Girder rail. The type of rail used with street railways having the head of the rail flush with the surface of the street.

Grand union. An intersection of two double-track trolley lines in which cars on both lines are enabled to turn either right or left.

Ground. The means by which spent DC voltage is returned to the power source, ordinarily the running rails.

Hackney coach (hack). An urban conveyance for hire; a taxicab.

Headway. The scheduled time on a given transit line between one vehicle and the next one.

High-level boarding. A station design wherein the platforms are at the same level as the floors of the cars.

Horsecar. A streetcar drawn either by a single horse (or mule) or a team.

Internal combustion. An engine in which the fuel burns inside the cylinder, ignited either by compression or by a spark, rather than in an external firebox, as with a steam engine.

Interurban. An electric railway connecting two or more cities or towns.

Jitney. (1) A transit vehicle, usually a large automobile, operated independently of municipal rules and regulations. (2) Also a nickel, typically the fare on a jitney.

Light rail. A system characterized by single cars or trains powered by overhead wires that operates along private rights-of-way or in the street, with boarding either at street level or platform level.

Light rail vehicle (LRV). Term used as a modernized equivalent of streetcar; *see* light rail.

Master Unit. A streetcar with standardized components developed by the J. G. Brill Co. in the 1920s.

Metro. A generic name for transit systems, and specifically for the heavy-rail system in Washington, D.C.

Monorail. A type of transit infrastructure from which the cars are either suspended or run along the top with guide-wheels to ensure stability.

Motorman. The operator of a streetcar.

Multiple unit control (MU). A system that enables the operator of the first (lead) car in a train of two or more cars to control the motors in each of the cars.

Near-side car. A car in which passengers board at a right-hand front door, pay a conductor stationed at the front, and normally exit through a left-hand front door. Near-side cars stop for passengers before crossing an intersection rather than on the far side of an intersection.

NIMBY (not in my back yard). Local or neighborhood resistance to the construction and operation of a transit line.

Omnibus. A horse-drawn conveyance operated over a fixed route at scheduled times of the day and with a set fare for all.

One-man operation. Operation of a streetcar by only the motorman or operator, who performs the conductor's duties as well.

Open car. A streetcar with transverse seats that is boarded from any point along the sides by means of a longitudinal step; *see also* California car.

Operator. The employee on a streetcar or LRV having direct control of the car's operation.

Pantograph. A hinged device with a flat contacting surface that picks up current from catenary, standard on electric streetcars in many parts of the world instead of trolley poles.

Pay-As-You-Enter (PAYE) car. A streetcar in which passengers board through a rear door, pay a conductor stationed at the rear, and usually exit front.

PCC. A streetcar developed through an industry-sponsored R&D effort, which was manufactured between 1936 and 1951, and became the last type remaining in operation in most American cities, although trams using PCC patents continued to be made in Europe into the 1980s.

People mover. A small-scale automated system of fixed-guideway transit.

Peter Witt car. A car in which passengers board in front and pay the conductor either as they step to the rear half of the car or exit at the center doors; the same front-entrance, center-exit (*see* treadle) design is typical of one-man operation.

Presidents' Conference Committee. *See* PCC.

Private right-of-way (PRW, *also* reserved ground, neutral ground, median). Streetcar tracks that are separated from mixed street traffic.

Rail (track). A rolled steel shape, commonly in a T-section, designed to be laid end-to-end on two parallel lines to support a conveyance with flanged wheels.

Running rail. The surface on which the tread of a flanged wheel bears.

Safety zone. An area in the street marked off by painted lines, or sometimes a raised platform, from which passengers board and alight from streetcars, ostensibly protected from automotive traffic.

Shuttle. A short streetcar line, usually with stops only at each end, often served by a single car.

Single ender. A trolley with controls only at one end.

Single trucker. A streetcar with only two axles.

St. Louis. The St. Louis Car Co., manufacturer of trolley and rapid-transit cars, 1887–1972.

Standard light rail vehicle (SLRV). An LRV designed and manufactured by Boeing Vertol with funding from the U.S. Department of Transportation; operated only in Boston and San Francisco.

Standee windows. Small oval windows in the sides of a streetcar or bus, at a height that enables standing passengers to see out.

Streetcar. A vehicle with flanged wheels and overhead trolley for collecting current, which operates in mixed street traffic.

Street running. The placement of trolley or LRV tracks so that operation is mixed with automobiles, buses, and trucks.

Substation. A structure containing rectifiers, breakers, and other equipment used to change power from a local electrical utility into power to be transmitted to an electric railway.

Subway (*also* Metro, the Underground, the Tube). An electrified heavy-rail line running mostly or entirely beneath the surface.

Surface transit. The operation of transit conveyances on the surface, usually meaning city streets, rather than overhead, under ground, or otherwise segregated from "mixed" traffic.

Transit mall. A downtown street or section of a central business district (CBD) where transit vehicles have priority or are given exclusive use.

Third rail. An electrified rail, usually mounted just outside the running rails, that conveys high-voltage current to "shoes" projecting from the trucks of rapid-transit trains.

Tie. The transverse members of a track structure to which the rails are spiked or bolted at the proper gauge in order to cushion and distribute the stresses of traffic through the ballast to the roadbed.

Timetable. A list indicating the times that transit conveyances are scheduled to depart from stops or stations.

Track. An assemblage of rails, ties, and fastenings over which rolling stock moves.

Trackless trolley (*also* trolley bus, trolley coach, or ETB). A rubber-tired conveyance with steering gear, powered by electric motors which derive current from a pair of trolley poles mounted on the roof (the second wire providing the return or ground).

Traction. The process involved when a conveyance applies its motive force directly through its own wheels (as with an electric trolley car) rather than being impelled by an external force.

Tram. The term used in most parts of the world other than North America for streetcar or trolley.

Transit authority. A public instrumentality charged with responsibility for operating mass transit.

Transit Research Corporation. The private entity that controlled patents for various components of PCC cars and licensed them for use.

Treadle. The device by which an alighting passenger triggers the opening of an exit door in the center or at the rear of a transit conveyance.

Troller. A current-collecting device with four grooved wheels designed to ride on top of paired overhead wires, one positive and the other negative.

Trolley. (1) Synonymous with streetcar or car. (2) The device on the roof of a streetcar that picks up current from an overhead wire by means of a pole with a small wheel or a carbon shoe.

Trolley bus. *See* trackless trolley.

Trolley pole. Pole atop a streetcar for collecting electrical power from an overhead wire; paired poles used with trackless trolleys.

Truck. An assemblage of components—frame, springs, axles, bearings, wheels, and usually gears and motors—upon which a streetcar rides, one, two, occasionally three of them per car.

Turnout. A point at which tracks branch into another route, or, with single track, on to a siding.

Urban Mass Transit Administration (UMTA, now Federal Transit Administration). The governmental agency established in 1964 and charged with granting financial assistance to transit authorities for capital improvements, and later, for subsidizing operation.

Bibliography

American Public Transit Association. www.apta.com.

Bail, Eli. *From Railway to Freeway: Pacific Electric and the Motorcoach*. Glendale, CA: Interurban Press, 1984.

Barrett, Paul. *The Automobile and Urban Transit*. Philadelphia, PA: Temple University Press, 1983.

Beebe, Lucius, and Charles Clegg. *Cable Car Carnival*. Oakland, CA: Grahame Hardy, 1951.

Bianco, Martha J. "The Decline of Transit: Corporate Conspiracy or Failure of Public Policy?: The Case of Portland, Oregon," *Journal of Policy History* 9 (Winter 1997).

Bobrick, Benson. *Labyrinths of Iron: Subways in History, Myth, Art, Technology, and War*. New York: William Morrow, 1981.

Bond, Winstan. "A Streetcar Named Success: The PCC, A Product of American Research in the 1930s," in *Perspectives on Railway History*, ed. Colin Divall. York, UK: Institute of Railway Studies, 1997.

Boston Elevated Railway Co. *Fifty Years of Unified Transportation in Metropolitan Boston*. Boston, MA: Boston Elevated Railway Co., 1938.

Bottles, Scott L. *Los Angeles and the Automobile: The Making of the Modern City*. Berkeley: University of California Press, 1987.

Brill, Debra. *History of the J. G. Brill Company*. Bloomington: Indiana University Press, 2001.

Charlton, E. Harper. *Railway Car Builders of the United States and Canada*. Los Angeles: Interurbans, 1957.

Cheape, Charles W. *Moving the Masses: Urban Public Transit in New York, Boston, and Philadelphia, 1880–1912*. Cambridge, MA: Harvard University Press, 1980.

Cox, Harold E. *PCC Cars of North America*. Philadelphia, PA: Author, 1963.

Cox, Harold E. *The Birney Car*. Forty Fort, PA: Author, 1966.

Crosby, Oscar T., and Louis Bell. *The Electric Railway in Theory and Practice*. New York: W. J. Johnston, 1893.

Cudahy, Brian J. *Change at Park Street Under: The Story of Boston's Subways*. Brattleboro, VT: Stephen Greene Press, 1972.

Cudahy, Brian J. *Under the Sidewalks of New York: The Story of the Greatest Subway System in the World*. Brattleboro, VT: Stephen Greene Press, 1979.

Cudahy, Brian J. *Cash, Tokens, and Transfers: A History of Urban Mass Transit in North America*. New York: Fordham University Press, 1995.

Diers, John. "Did a Conspiracy Really Kill the Streetcar?" *Trains* 66 (January 2006).

Divall, Colin, and Winstan Bond, eds. *Suburbanizing the Masses: Public Transport and Urban Development in Historical Perspective*. Aldershot, UK: Ashgate, 2003.

Duke, Donald, ed. *Denver and Interurban (Fort Collins Division) and the Fort Collins Municipal Railway*. San Marino, CA: Pacific Railway Journal, 1957.

Emmons, Charles D. *The Development of American Street Railways*. New York: American Electric Railway Association, 1924.

Fairchild, C. B. *Street Railways: Their Construction, Operation, and Maintenance*. New York: Street Railway Publishing, 1892.

Farrell, Michael R. *Who Made All Our Streetcars Go? The Story of Rail Transit in Baltimore*. Baltimore, MD: Baltimore National Railway Historical Society Publications, 1973.

Flink, James J. *The Automobile Age*. Cambridge, MA: MIT Press, 1988.

Foster, Mark S. *From Streetcar to Superhighway: American City Planners and Urban Transportation, 1900–1940*. Philadelphia, PA: Temple University Press, 1981.

Glaab, Charles N., and A. Theodore Brown. *A History of Urban America*. London: Collier-Macmillan Ltd., 1967.

Hall, Sir Peter. *Cities in Civilization*. New York: Pantheon Books, 1998.

Hallidie, Andrew S. *A Brief History of the Cable Railway System*. San Francisco: n.p., 1891.

Hallidie, Andrew. *The Invention of the Cable Railway System*. San Francisco: n.p., 1885.

Hamm, Edward, Jr. *The Public Service Trolley Lines of New Jersey*. Polo, IL: Transportation Trails, 1991.

Hanscom, W. W. *Cable Railway Propulsion*. San Francisco: Technical Society of the Pacific Coast, 1884.

Harwood, Herbert H. *Baltimore Streetcars: The Postwar Years*, rev. ed. Baltimore, MD: Johns Hopkins University Press, 2003.

Hering, Carl. *Recent Progress in Electric Railways*. New York: W. J. Johnston Co., 1892.

Herrick, Albert B., and Edward C. Boynton. *American Electric Railway Practice*. New York: McGraw, 1907.

Hilton, George W. *The Cable Car in America*, rev. ed. San Diego, CA: Howell-North, 1982.

Hilton, George W., and John F. Due. *The Electric Interurban Railways in America*. Stanford, CA: Stanford University Press, 1960.

Hood, Clifton. *772 Miles: The Building of the Subways and How They Transformed New York*. Baltimore, MD: Johns Hopkins University Press, 1993.

Jackson, Kenneth T. *Crabgrass Frontier: The Suburbanization of the United States*. New York: Oxford University Press, 1985.

Jones, David W., Jr. *Urban Transit Policy: An Economic and Political History*. Englewood Cliffs, NJ: Prentice-Hall, 1985.

Kahn, Edgar M. *Cable Car Days in San Francisco*. Stanford, CA: Stanford University Press, 1940.

Kashin, Seymour, and Harre Demoro. *The PCC Car: An American Original*. Glendale, CA: Interurban Press, 1986.

Kieffer, Stephen A. *Transit and the Twins*. Minneapolis, MN: Twin City Rapid Transit Co., 1958.

King, LeRoy O. *100 Years of Capital Traction: The Story of Streetcars in the Nation's Capital*. Dallas, TX: Author, 1972.

Klapper, Charles. *The Golden Age of Tramways*. North Pomfret, VT: David & Charles, 1974.

Luke, William A., and Linda L. Metler. *City Transit Buses of the Twentieth Century*. Hudson, WI: Iconografix, 2000.

McKay, John P. *Tramways and Trolleys: The Rise of Urban Mass Transit in Europe*. Princeton, NJ: Princeton University Press, 1976.

McKenney, Carlton N. *Rails in Richmond*. Glendale, CA: Interurban Press, 1986.

McShane, Clay. *Down the Asphalt Path: The Automobile and the American City*. New York: Columbia University Press, 1994.

Meyers, Stephen L. *Manhattan's Lost Streetcars*. Chicago, IL: Arcadia Publishing, 2005.

Middleton, William D. *The Time of the Trolley: The Street Railway from Horsecar to Light Rail*. San Marino, CA: Golden West Books, 1987.

Miller, John Anderson. *Fares Please! From Horse-cars to Streamliners*. New York: D. Appleton-Century Co., 1943.

Mom, Gijs. "Roads and Rails: European Highway-Network Building and the Desire for Long-Range Motorized Mobility," *Technology and Culture* 46 (2005).

National Cable Railway Co. *The System of Wire-Cable Railways for Cities and Towns*. New York: National Cable Railway Co., 1887.

Passer, Harold W. *The Electrical Manufacturers, 1875–1900: A Study in Competition, Entrepreneurship, Technical Change, and Economic Growth*. Cambridge, MA: Harvard University Press, 1953.

Pattison, Tony. *Jane's Urban Transport Systems, Twentieth Edition 2001–2002*. Coulsdon, UK: Jane's Information Group Ltd., Sentinel House, 2001.

Plous, F. K., Jr. "A Desire Named Streetcar," *Planning*, June 1984.

Post, Robert C. "The Fair Fare Fight: An Episode in Los Angeles History," *Southern California Quarterly* 52 (1970).

Post, Robert C. "Electromagnetism as a Motive Power: Robert Davidson's *Galvani*," *Railroad History* 130 (Spring 1974).

Post, Robert C. *Physics, Patents, and Politics: A Biography of Charles Grafton Page*. New York: Science History Publications, 1976.

Post, Robert C. "America's Electric Railway Beginnings: Trollers and Daft Dummies in Los Angeles," *Southern California Quarterly* 69 (1987).

Post, Robert C. *Street Railways and the Growth of Los Angeles*. San Marino, CA: Golden West Books, 1989.

Post, Robert C. "Renaissance Man: An Interview with George Krambles," *Railroad History* 175 (Autumn 1996).

Post, Robert C. "The Myth behind the Streetcar Revival," *American Heritage* 49 (May/June 1998).

Pye, David. *The Nature and Aesthetics of Design*. London: Barrie & Jenkins, 1978.

Rae, John B. *The American Automobile*. Chicago, IL: University of Chicago Press, 1965.

Reed, Robert C. *The New York Elevated*. New York: A. S. Barnes & Co., 1978.

Richmond, Jonathan. *Transport of Delight: The Mythical Conception of Rail Transit in Los Angeles*. Akron, OH: University of Akron Press, 2005.

Richter, William L. *Transportation in America*. Santa Barbara, CA: ABC-Clio, 1995.

Rowsome, Frank. *Trolley Car Treasury: A Century of American Streetcars*. New York: Bonanza Books, 1956.

Scharff, Virginia. *Taking the Wheel: Women and the Coming of the Motor Age*. New York: The Free Press, 1991.

Schatzberg, Eric. "Culture and Technology in the City: Opposition to Mechanized Street Transportation in Late-Nineteenth-Century America," in *Technologies of Power*, ed. Michael Thad Allen and Gabrielle Hecht. Cambridge, MA: MIT Press, 2001.

Schivelbusch, Wolfgang. *The Railway Journey: Trains and Travel in the 19th Century*. New York: Urizen Books, 1979.

Schneider, Fred W., and Stephen P. Carlson. *PCC from Coast to Coast*. Glendale, CA: Interurban Press, 1983.

Schrag, Zachary M. "'The Bus Is Young and Honest': Transportation Politics, Technical Choice, and the Motorization of Manhattan Surface Transit, 1919–1936," *Technology and Culture* 41 (2000).

Schrag, Zachary M. *The Great Society Subway: A History of the Washington Metro*. Baltimore, MD: Johns Hopkins University Press, 2006.

Sebree, Mac, and Paul Ward. *Transit's Stepchild: The Trolley Coach*. Cerritos, CA: Interurbans, 1973.

Slater, Cliff. "General Motors and the Demise of Streetcars," *Transportation Quarterly* 51 (Summer 1997).

Smith, J. Bucknall. *A Treatise on Rope or Cable Traction*. London: Engineering, 1887.

Snell, Bradford C., for the Senate Judiciary Committee. *American Ground Transport: A Proposal for Restructuring the Automobile, Truck, Bus, and Rail Industries*. Washington, DC: Government Printing Office, 1974.

Sprague, Frank J. "The Story of the Trolley Car," *The Century Magazine* 70 (July/August 1905).

St. Clair, David J. *The Motorization of American Cities*. New York: Praeger, 1986.

Stevens, John R. *Pioneers of Electric Railroading*. New York: Electric Railroaders' Assn., 1991.

Swett, Ira L. *Los Angeles Railway*. Los Angeles: Interurbans, 1951.

Taylor, George Rogers. "The Beginnings of Mass Transportation in Urban America," *The Smithsonian Journal of History* 1 (Summer/Autumn 1966).

Thorpe, James. *Henry Edwards Huntington: A Biography*. Berkeley: University of California Press, 1994.

Todd, Sarah. *The History of Urban Transport*. York, UK: Institute of Railway Studies, 1999.

Toman, James A., and Blaine S. Hays. *Horse Trails to Regional Rails: The Story of Public Transit in Greater Cleveland*. Kent, OH: Kent State University Press, 1996.

Vincenti, Walter G. *What Engineers Know and How They Know It: Analytical Studies from Aeronautical History*. Baltimore, MD: Johns Hopkins University Press, 1990.

Volti, Rudi, ed. *The Facts on File Encyclopedia of Science, Technology, and Society*. New York: Facts on File, 1999.

Volti, Rudi. *Cars and Culture: The Life Story of a Technology*. Westport, CT: Greenwood Press, 2004.

Warner, Sam B., Jr. *Streetcar Suburbs: The Process of Growth in Boston, 1870–1900*. Cambridge, MA: Harvard University Press, 1962.

Wattenberg, Ben J. *The Statistical History of the United States from Colonial Times to the Present*. New York: Basic Books, 1976.

Whalen, Grover. *Replacing Street Cars with Motor Buses*. New York: National Automobile Chamber of Commerce, 1920.

Whipple, Fred H. *The Electric Railway*. Detroit, MI: William Graham, 1889; reprint ed. 1980.

White, John H., Jr., ed. *Horsecars, Cable Cars and Omnibuses: All 107 Photographs from the John Stephenson Company Album, 1888*. New York: Dover Publications, 1974.

White, John H., Jr. "Spunky Little Devils: Locomotives of the New York Elevated," *Railroad History* 162 (Spring 1990).

White, John H., Jr. "War of the Wires: A Curious Chapter in Street Railway History," *Technology and Culture* 46 (2005).

Williams, Rosalind. *Notes on the Underground: An Essay on Technology, Society, and the Imagination*. Cambridge, MA: MIT Press, 1990 .

Wood, Donald F. *American Buses*. Osceola, WI: MBI Publishing, 1998.

Wright, Augustine W. *American Street Railways*. Chicago, IL: Rand McNally & Co, 1888.

Yago, Glenn. *The Decline of Transit: Urban Transportation in German and U.S. Cities, 1900–1970*. Cambridge: Cambridge University Press, 1984.

Young, Andrew D. *St. Louis Car Company Album*. Glendale, CA: Interurban Press, 1984.

Young, David M. *Chicago Transit: An Illustrated History*. DeKalb, IL: Northern Illinois University Press, 1998.

Index

About the Author

ROBERT C. POST received his doctorate in American history from the University of California, Los Angeles. From 1973 to 1996 he was employed by the Smithsonian Institution, first by the National Museum of History and Technology, then by Smithsonian Books, and finally by the National Museum of American History. His books include *Physics, Patents, and Politics: A Biography of Charles Grafton Page* (1976), *Street Railways and the Growth of Los Angeles* (1989), and *Technology, Transport, and Travel in American History* (2003). For fifteen years he was editor of *Technology and Culture*, the quarterly journal of the Society for the History of Technology (SHOT). He was SHOT's president in 1997–98 and recipient of its Leonardo Da Vinci Medal in 2001.